HUMAN MICROBIOME

HOW BACTERIA IN THE BODY INFLUENCE OUR HEALTH

DAVID SANDUA

Human Microbiome.
© David Sandua 2024. All rights reserved.
Electronic and paperback edition.

"Research on the human microbiome teaches us that we are more than human cells, we are communities of microorganisms in symbiosis with us."

María Gloria Domínguez-Bello

INDEX

I. INTRODUCTION 13
  Definition of the Human Microbiome 13
  Importance of Microbiome Research 14
  Thesis Statement: Exploring the Impact of Bacteria on Health 15

II. HISTORICAL PERSPECTIVE ON MICROBIOME STUDIES 16
  Early Discoveries and Theories 16
  Advances in Microbial Research 17
  The Shift Towards Microbiome Analysis 17

III. COMPOSITION OF THE HUMAN MICROBIOME 19
  Types of Microorganisms Present 19
  Variability Across Different Body Sites 20
  Factors Influencing Microbiome Composition 21

IV. METHODOLOGIES FOR STUDYING THE MICROBIOME 22
  DNA Sequencing Technologies 22
  Metagenomics and Bioinformatics 23
  Challenges in Microbiome Research 24

V. THE GUT MICROBIOME AND DIGESTIVE HEALTH 25
  Role in Digestion and Nutrient Absorption 25
  Influence on Gastrointestinal Diseases 26
  Probiotics and Gut Health 26

VI. MICROBIOME AND IMMUNE SYSTEM INTERACTION 28
  Development of Immune System 28
  Microbiome's Role in Immune Modulation 29
  Impact on Autoimmune Diseases 30

VII. THE SKIN MICROBIOME 31
  Composition and Function 31
  Relationship with Skin Conditions 32
  Therapeutic Approaches for Skin Health 32

VIII. THE ORAL MICROBIOME 34
  Composition and Its Unique Environment 34
  Link to Oral Diseases 35
  Preventive Measures and Oral Health 35

IX. THE RESPIRATORY MICROBIOME 37
  Microbial Communities in the Respiratory Tract 37
  Impact on Respiratory Health 38
  Future Directions in Respiratory Microbiome Research 38

X. THE UROGENITAL MICROBIOME 40
  Composition and Health Implications 40
  Influence on Reproductive Health 41
  Strategies for Managing Urogenital Health 41

XI. MICROBIOME AND METABOLIC SYNDROME 43
  Influence on Obesity and Diabetes 43
  Mechanisms Linking Microbiome to Metabolism 44
  Intervention Strategies 44

XII. MICROBIOME AND CARDIOVASCULAR HEALTH 46
IMPACT ON HEART DISEASE 46
MECHANISMS OF INFLUENCE 47
POTENTIAL FOR THERAPEUTIC INTERVENTIONS 48

XIII. MICROBIOME AND MENTAL HEALTH 49
CONCEPT OF THE GUT-BRAIN AXIS 49
MICROBIOME'S ROLE IN MENTAL DISORDERS 50
PROBIOTICS AND MENTAL HEALTH TREATMENTS 50

XIV. MICROBIOME AND CANCER 52
INFLUENCE ON CANCER DEVELOPMENT 52
MICROBIOME AS A DIAGNOSTIC TOOL 53
MICROBIOME-TARGETED THERAPIES 54

XV. MICROBIOME AND AGING 55
CHANGES IN MICROBIOME OVER LIFE SPAN 55
IMPACT ON AGE-RELATED DISEASES 56
POTENTIAL INTERVENTIONS TO PROMOTE HEALTHY AGING 56

XVI. PEDIATRIC MICROBIOME AND DEVELOPMENT 58
ESTABLISHMENT OF THE MICROBIOME IN INFANCY 58
IMPACT ON CHILD DEVELOPMENT AND HEALTH 59
STRATEGIES FOR OPTIMIZING PEDIATRIC HEALTH 59

XVII. DIET AND THE MICROBIOME 61
EFFECTS OF DIETARY CHOICES 61
DIET-BASED MODULATION OF THE MICROBIOME 62
RECOMMENDATIONS FOR MICROBIOME-FRIENDLY DIETS 62

XVIII. ANTIBIOTICS AND THE MICROBIOME 64
IMPACT OF ANTIBIOTIC USE 64
STRATEGIES TO MITIGATE NEGATIVE EFFECTS 65
FUTURE OF ANTIBIOTIC POLICIES 65

XIX. FECAL MICROBIOTA TRANSPLANTATION (FMT) 67
PRINCIPLES AND PROCEDURES 67
CLINICAL APPLICATIONS 68
ETHICAL AND REGULATORY CONSIDERATIONS 69

XX. ETHICAL CONSIDERATIONS IN MICROBIOME RESEARCH 70
PRIVACY AND DATA MANAGEMENT 70
CONSENT AND PARTICIPATION 71
IMPLICATIONS OF MICROBIOME MANIPULATION 71

XXI. PUBLIC AWARENESS AND EDUCATION ON THE MICROBIOME 73
CURRENT PUBLIC KNOWLEDGE LEVELS 73
IMPORTANCE OF EDUCATING THE PUBLIC 74
STRATEGIES FOR EFFECTIVE COMMUNICATION 74

XXII. REGULATORY CHALLENGES IN MICROBIOME RESEARCH 76
OVERVIEW OF REGULATORY LANDSCAPE 76
CHALLENGES IN STANDARDIZING PROTOCOLS 77
FUTURE DIRECTIONS IN REGULATION 78

XXIII. GLOBAL VARIATIONS IN HUMAN MICROBIOMES 79
GEOGRAPHIC AND CULTURAL DIFFERENCES 79
IMPLICATIONS FOR GLOBAL HEALTH 80
STRATEGIES FOR CROSS-CULTURAL RESEARCH 81

XXIV. THE ROLE OF GENETICS IN THE MICROBIOME 82
- Genetic Influences on Microbiome Composition 82
- Personalized Medicine Approaches 83
- Future Research Directions in Genomics and Microbiomes 84

XXV. TECHNOLOGICAL INNOVATIONS IN MICROBIOME RESEARCH 85
- New Tools and Techniques 85
- Impact on Research Efficiency and Accuracy 86
- Future Technological Trends 87

XXVI. FUTURE THERAPEUTIC POTENTIALS OF THE MICROBIOME 88
- Emerging Therapeutic Techniques 88
- Challenges in Therapeutic Application 89
- Predictions for Future Therapies 90

XXVII. MICROBIOME AND PERSONALIZED MEDICINE 91
- Tailoring Treatments Based on Microbiome 91
- Challenges in Implementing Personalized Approaches 92
- Future of Personalized Medicine and Microbiome 93

XXVIII. MICROBIOME IN NON-HUMAN MODELS 94
- Studies on Animal Microbiomes 94
- Ethical Considerations in Animal Studies 96

XXIX. COMMERCIALIZATION OF MICROBIOME RESEARCH 97
- Current Market Trends 97
- Ethical and Practical Challenges 98
- Future Market Predictions 99

XXX. PARTNERSHIPS AND COLLABORATIONS IN MICROBIOME RESEARCH 100
- Role of Interdisciplinary Collaboration 100
- Major Collaborative Projects 101
- Benefits and Challenges of Collaboration 101

XXXI. FUNDING AND INVESTMENT IN MICROBIOME RESEARCH 103
- Overview of Funding Sources 103
- Trends in Investment 104
- Impact of Funding on Research Progress 105

XXXII. MICROBIOME RESEARCH AND PUBLIC HEALTH POLICY 106
- Influence on Health Policy Making 106
- Policy Challenges 107
- Recommendations for Policymakers 108

XXXIII. MICROBIOME AND ENVIRONMENTAL HEALTH 109
- Interaction Between Environmental Factors and Microbiome 109
- Impact on Public Health 110
- Strategies for Environmental Management 111

XXXIV. CHALLENGES IN MICROBIOME SAMPLE COLLECTION AND STORAGE 112
- Best Practices for Sample Collection 112
- Storage and Preservation Techniques 113
- Impact on Research Quality 113

XXXV. DATA ANALYSIS AND INTERPRETATION IN MICROBIOME RESEARCH 115
- Advanced Analytical Techniques 115
- Challenges in Data Interpretation 116
- Improving Accuracy and Reliability 117

XXXVI. MICROBIOME AND INFECTIOUS DISEASES 118

- Role in Disease Prevention 118
- Microbiome's Influence on Pathogen Dynamics 119
- Strategies for Infectious Disease Management 119

XXXVII. MICROBIOME AND RESISTANCE TO ANTIBIOTICS 121
- Development of Resistance 121
- Strategies to Combat Resistance 122
- Future Directions in Research and Treatment 122

XXXVIII. LEGAL ASPECTS OF MICROBIOME RESEARCH 124
- Intellectual Property Issues 124
- Compliance with International Laws 125
- Future Legal Challenges 125

XXXIX. MICROBIOME AND LIFESTYLE FACTORS 127
- Impact of Exercise on Microbiome 127
- Effects of Stress and Sleep 128
- Lifestyle Modifications for Optimal Microbiome Health 128

XL. ROLE OF MICROBIOME IN NUTRACEUTICALS 130
- Microbiome-targeted Nutraceuticals 130
- Efficacy and Safety Considerations 131
- Market and Regulatory Aspects 131

XLI. MICROBIOME AND VETERINARY MEDICINE 133
- Applications in Animal Health 133
- Comparative Studies with Human Microbiomes 134
- Future Directions in Veterinary Applications 135

XLII. MICROBIOME AND AGRICULTURAL SCIENCES 136
- Impact on Soil and Plant Health 136
- Applications in Sustainable Agriculture 137
- Future Agricultural Strategies 137

XLIII. MICROBIOME AND FOOD INDUSTRY 139
- Influence on Food Processing 139
- Probiotics in Food Products 140
- Future Trends in Food Technology 140

XLIV. GLOBAL HEALTH INITIATIVES AND THE MICROBIOME 142
- International Health Programs 142
- Role of Microbiome Research in Global Health 143
- Strategies for Global Health Improvement 144

XLV. MICROBIOME AND BIOTECHNOLOGY 145
- Biotechnological Applications 145
- Innovations in Microbiome Engineering 146
- Ethical and Safety Considerations 147

XLVI. CHALLENGES IN TRANSLATING MICROBIOME RESEARCH 148
- From Laboratory to Clinic 148
- Barriers in Clinical Application 149
- Strategies for Overcoming Challenges 150

XLVII. MICROBIOME AND PUBLIC SAFETY 151
- Biosecurity Concerns 151
- Microbiome in Disease Surveillance 152
- Public Safety Strategies 152

XLVIII. FUTURE DIRECTIONS IN MICROBIOME RESEARCH 154

Emerging Research Areas 154
　　Potential Breakthroughs 155
　　Long-term Research Goals 156

XLIX. SUMMARY OF KEY FINDINGS 157
　　Major Insights from the Essay 157
　　Implications for Future Research 158
　　Relevance to Health and Disease 158

L. IMPLICATIONS FOR POLICY AND PRACTICE 160
　　Recommendations for Health Practitioners 160
　　Policy Implications 161
　　Practical Applications of Research Findings 161

LI. CONCLUSION 163
　　Recapitulation of Thesis and Main Points 163
　　Future Outlook in Microbiome Research 164
　　Closing Remarks 165

BIBLIOGRAPHY 166

I. INTRODUCTION

The intricate relationship between the human microbiome and our health has garnered increasing attention in the field of medicine and biology. The human body is host to trillions of microorganisms, primarily bacteria, that coexist in a delicate balance with our cells. The composition of our microbiome can vary significantly from person to person, influenced by factors such as diet, genetics, and environment. These microbiota play a fundamental role in shaping our immune system, metabolism, and even mental health. Understanding the dynamic interplay between the human microbiome and our health is crucial for developing personalized medicine approaches and innovative treatments for a wide range of diseases. As we delve deeper into this microbial world within us, we uncover new insights into how bacteria influence our physiology and hold the key to unlocking groundbreaking advancements in healthcare.

Definition of the Human Microbiome

The human microbiome, defined as the collection of microorganisms residing in and on the human body, is a complex ecosystem that significantly influences our health. These microbes, including bacteria, viruses, and fungi, play a crucial role in various physiological processes, from nutrient absorption to immune system function. The composition of the human microbiome is unique to each individual and can be influenced by factors such as genetics, diet, and environment. Recent advancements in technology have allowed researchers to better understand the diversity and functions of these microbial communities, shedding light on their impact on human health. By studying the human microbiome, scientists can gain insights into diseases such as obesity, autoimmune disorders, and even mental health conditions. Understanding the intricate relationship between our bodies and the microorganisms that inhabit them is essential for developing personalized strategies to promote overall well-being.

Importance of Microbiome Research

Recent advancements in microbiome research have shed light on the crucial role that bacteria in our bodies play in influencing our health. Understanding the complex interactions within the microbiome can provide valuable insights into various health conditions, ranging from gastrointestinal disorders to mental health issues. By examining the composition of microbial communities and their impact on physiological processes, researchers can develop targeted interventions and personalized treatments. Moreover, microbiome research has the potential to revolutionize healthcare by providing a new perspective on disease prevention and management. Through the identification of key microbial players and their functions, researchers aim to uncover novel therapeutic strategies that can help improve patient outcomes and enhance overall well-being. Ultimately, the importance of microbiome research lies in its ability to elucidate the intricate relationship between our bodies and the trillions of microorganisms that inhabit us, offering promising avenues for advancing medical science.

Thesis Statement: Exploring the Impact of Bacteria on Health

Recent research has shed light on the intricate relationship between bacteria and human health, revealing the significant impact these microorganisms have on our well-being. By colonizing our gut, skin, and other bodily systems, bacteria play a crucial role in maintaining a delicate balance that is essential for optimal health. The human microbiome, which consists of trillions of microbes, interacts with our immune system, influences nutrient absorption, and even affects our mental health. Dysbiosis, or an imbalance in the microbiome, has been linked to a range of health conditions, including obesity, autoimmune diseases, and mental health disorders. Understanding the complex interplay between bacteria and health is essential for developing targeted interventions to promote overall well-being. As we delve deeper into this complex ecosystem of microbes, we continue to uncover the profound ways in which bacteria impact our health and the potential for harnessing this knowledge to improve health outcomes for individuals around the world.

II. HISTORICAL PERSPECTIVE ON MICROBIOME STUDIES

As microbiome studies have gained traction in recent years, it is important to consider the historical perspective that has shaped this field of research. The roots of microbiome studies can be traced back to the work of Antonie van Leeuwenhoek, who first observed microscopic organisms in the 17th century. However, it was not until the late 20th century that advancements in DNA sequencing technology allowed researchers to delve deeper into the complexities of the human microbiome. This technological breakthrough enabled scientists to identify and characterize the vast array of microorganisms living in and on the human body. By examining the historical progression of microbiome studies, we can appreciate the evolution of our understanding of the intricate relationship between the microbiome and human health. This historical context provides valuable insights into the current state of microbiome research and underscores the importance of interdisciplinary collaboration in unraveling the mysteries of the microbiome.

Early Discoveries and Theories
The early discoveries and theories in the field of human microbiome research have laid the foundation for our current understanding of how bacteria in the body influence our health. From Antonie van Leeuwenhoek's pioneering observations of microorganisms in the 17th century to Louis Pasteur's germ theory of disease in the 19th century, the idea that microorganisms play a significant role in human health has been a constant theme throughout the history of science. Over time, researchers have

uncovered the intricate interactions between the human host and the trillions of bacteria that reside within us, leading to groundbreaking discoveries about the microbiome's impact on various aspects of health, including digestion, immunity, and even mental health. These early findings have shaped the way we approach healthcare today, emphasizing the importance of maintaining a balanced microbiome for overall well-being.

Advances in Microbial Research

Recent advances in microbial research have shed light on the intricate relationship between the human microbiome and various health conditions. By utilizing cutting-edge techniques such as metagenomics and metabolomics, scientists have been able to uncover the vast diversity of microorganisms that reside in and on our bodies. These studies have revealed the crucial role that gut bacteria play in digestion, immune function, and even mental health. Understanding the composition and function of the human microbiome has paved the way for innovative therapies, such as fecal microbiota transplants, that harness the power of beneficial bacteria to treat conditions like Clostridioides difficile infection. Moreover, insights from microbial research have highlighted the importance of maintaining a diverse and balanced microbiome through diet, lifestyle, and the judicious use of antibiotics. As our knowledge of the microbiome continues to evolve, so too does our ability to harness its potential for optimizing health and well-being.

The Shift Towards Microbiome Analysis

As advancements in technology continue to evolve, there has been a noticeable shift towards microbiome analysis in the field of health and medicine. This shift is driven by the recognition of the significant impact that the microbiome has on human health. Researchers are increasingly interested in studying the complex interactions between the trillions of microorganisms that reside in and on the human body and how they influence various physiological processes. The use of techniques such as metagenomics and metabolomics allows for a more in-depth understanding of the intricate relationships between the host and its microbial inhabitants. By uncovering the nuances of the microbiome, scientists aim to identify potential biomarkers for certain diseases, develop personalized treatment strategies, and ultimately improve overall health outcomes for individuals. This emphasis on microbiome analysis marks a pivotal moment in the pursuit of precision medicine and underscores the importance of considering the microbiome as an integral component of human health.

III. COMPOSITION OF THE HUMAN MICROBIOME

Our understanding of the composition of the human microbiome has significantly evolved in recent years, revealing a complex ecosystem of bacteria, viruses, fungi, and other microorganisms that inhabit various parts of our body. The human microbiome is highly diverse, with distinct communities residing in different regions such as the skin, gut, mouth, and reproductive tract. The gut microbiota, in particular, is recognized for its vast array of species and crucial functions in digestion, metabolism, and immune regulation. Research indicates that the composition of the microbiome is influenced by various factors including diet, lifestyle, genetics, and environment. Understanding the specific composition of an individual's microbiome can provide valuable insights into their health status and susceptibility to certain diseases. Moreover, advances in sequencing technologies have enabled researchers to characterize the microbiome at a much deeper level, shedding light on the intricate interactions between microbial communities and the host. Such insights hold great promise for personalized medicine and innovative therapeutic interventions aimed at restoring microbial balance and promoting overall well-being.

Types of Microorganisms Present
The human microbiome is a complex ecosystem composed of various types of microorganisms that reside in different areas of the body. These microorganisms include bacteria, viruses, fungi, and archaea, with bacteria being the most abundant and diverse group. Among the bacteria, there are different species and

strains that vary in their functions and effects on human health. For instance, some bacteria are considered beneficial or probiotic, aiding in digestion, immune function, and vitamin synthesis, while others can be pathogenic and cause infections or disease. Understanding the types of microorganisms present in the human microbiome is crucial in elucidating their roles in maintaining health and preventing illness. By studying the interactions between these microorganisms and the human body, researchers can develop strategies to promote a healthy microbiome and improve overall well-being. Thus, identifying and characterizing the diverse range of microorganisms present in the human microbiome is essential for advancing our knowledge of how these organisms influence our health.

Variability Across Different Body Sites
Moreover, the variability across different body sites in terms of microbial composition adds another layer of complexity to the human microbiome. Research has shown that the microbiota present in the gut differ significantly from those on the skin or in the oral cavity. This diversity is crucial as different body sites have unique environmental conditions that select for specific microbial communities. For example, the gut provides a
warm, moist, and nutrient-rich environment ideal for the growth of certain bacteria, while the skin offers a drier and more acidic habitat that favors a different set of microorganisms. Understanding the variations in microbial populations across various body sites is essential for deciphering the intricate relationship between the microbiome and its host. By exploring these differences, researchers can gain invaluable insights into how the microbiome influences human health and disease on a site-specific

level. This highlights the importance of considering the microbiome as a dynamic ecosystem with diverse niches that interact in a complex network within the human body.

Factors Influencing Microbiome Composition

The composition of the human microbiome is influenced by a multitude of factors, ranging from genetics to environmental exposures. One crucial determinant is the individual's diet, as certain foods can promote the growth of beneficial microbes while inhibiting pathogenic strains. Additionally, lifestyle choices such as exercise and stress management can also impact the balance of the microbiome. Furthermore, medications, such as antibiotics, can drastically alter the microbial community within the gut, leading to potential dysbiosis. Moreover, age and geographical location have been shown to influence microbiome composition, with infants having a different microbiome than adults and individuals from different regions exhibiting varied microbial populations. Understanding these factors is essential in order to maintain a healthy microbiome and prevent diseases associated with dysbiosis. Ultimately, the intricate interplay of these elements highlights the complexity of microbiome composition and its relevance to overall health.

IV. METHODOLOGIES FOR STUDYING THE MICROBIOME

Recent advances in technology have revolutionized the methodologies for studying the microbiome, enabling researchers to delve deeper into the intricate world of microbial communities within the human body. Metagenomic sequencing has emerged as a powerful tool, allowing for the comprehensive analysis of microbial DNA present in various body sites. This approach not only provides insights into the diversity and composition of the microbiome but also helps identify potential correlations between microbial profiles and health outcomes. Additionally, metabolomic analysis offers valuable information on the functional capabilities of these microbial communities, shedding light on their role in human physiology and disease. Integration of multi-omics data sets further enhances our understanding of the complex interactions within the microbiome. These cutting-edge methodologies hold great promise in unraveling the impact of the microbiome on human health and pave the way for personalized approaches to microbiome-based interventions.

DNA Sequencing Technologies

As advancements in technology have transformed the field of genomics, DNA sequencing technologies have played a pivotal role in unraveling the complexities of the human microbiome. High-throughput sequencing methods, such as next-generation sequencing (NGS), have revolutionized the way researchers examine microbial communities within the body. By providing a comprehensive view of the genetic material present in these communities, NGS allows for a deeper understanding of the diversity and functions of these microorganisms. Additionally, the development of long-read sequencing technologies has enabled the assembly of complete genomes from complex microbial populations, leading to insights into the functional potential of these microbes. These advancements in DNA sequencing have not only expanded our knowledge of the human microbiome but have also paved the way for personalized medicine approaches that harness the microbial composition of individuals for improved health outcomes.

Metagenomics and Bioinformatics

Metagenomics, a powerful tool in studying microbial communities, provides new insights into the complex interactions between bacteria and their human hosts. By sequencing the DNA of all microbes in a particular environment, metagenomics allows researchers to identify the diverse species present and their respective functions. Bioinformatics, on the other hand, plays a crucial role in analyzing and interpreting the vast amount of data generated through metagenomic studies. Through advanced computational techniques, bioinformatics helps researchers unravel the genetic code of bacteria and understand how they influence human health. By combining metagenomics and bioinformatics, scientists can uncover novel ways in which the human microbiome impacts various physiological processes. This interdisciplinary approach not only enhances our understanding of microbial communities but also opens up new avenues for personalized medicine and targeted interventions to promote better health outcomes.

Challenges in Microbiome Research

When diving into the realm of microbiome research, various challenges present themselves, hindering progress in understanding the intricate relationship between the human body and its resident bacteria. One significant challenge is the vast complexity and diversity of microbial communities inhabiting different parts of the body, making it difficult to study them comprehensively. Additionally, the dynamic nature of the microbiome, affected by various factors such as diet, lifestyle, and environmental exposures, poses a challenge in maintaining consistent research results. Furthermore, the lack of standardized methods for studying the microbiome hinders the comparability and reproducibility of studies across different research groups. Addressing these challenges requires interdisciplinary collaborations, advanced technological tools, and robust study designs to unravel the mysteries of the microbiome and its impact on human health. Despite these obstacles, the field of microbiome research holds immense potential for revolutionizing our understanding of health and disease.

V. THE GUT MICROBIOME AND DIGESTIVE HEALTH

Recent research has highlighted the intricate relationship between the gut microbiome and digestive health. The gut microbiota, composed of trillions of microorganisms, plays a vital role in maintaining gastrointestinal function, nutrient absorption, and immune response. Studies have shown that disruptions in the gut microbiome, known as dysbiosis, are associated with various digestive disorders such as irritable bowel syndrome (IBS), inflammatory bowel disease (IBD), and even colon cancer. The balance of beneficial and harmful bacteria in the gut influences the integrity of the intestinal barrier, inflammation levels, and the production of essential nutrients. Furthermore, the gut-brain axis demonstrates the bidirectional communication between the gut microbiome and the central nervous system, impacting mood, cognition, and even behavior. Understanding the dynamics of the gut microbiome is crucial for developing targeted interventions to promote digestive health and overall well-being.

Role in Digestion and Nutrient Absorption
The human microbiome plays a fundamental role in digestion and nutrient absorption, influencing various aspects of our overall health. Within the gastrointestinal tract, trillions of bacteria aid in breaking down food particles and extracting essential nutrients for absorption into the bloodstream. These microorganisms also have the ability to metabolize certain compounds that our bodies are unable to digest independently, such as fiber. By fermenting these indigestible substances, gut bacteria produce

short-chain fatty acids that not only serve as a crucial energy source for intestinal cells but also have anti-inflammatory properties. Moreover, the composition of the gut microbiota can affect the efficiency of nutrient absorption, potentially impacting nutrient deficiencies or excesses in the body. Therefore, understanding the intricate relationship between the human microbiome and digestion is vital in maintaining optimal health and well-being.

Influence on Gastrointestinal Diseases

Recent research has shown a clear link between the human microbiome and gastrointestinal diseases. The composition of the gut microbiota can influence the development and progression of conditions such as inflammatory bowel disease (IBD), irritable bowel syndrome (IBS), and even colorectal cancer. Dysbiosis, or an imbalance in the gut microbiota, has been implicated in the pathogenesis of these diseases, leading to chronic inflammation and compromised gut barrier function. Additionally, the gut microbiome plays a key role in the production of metabolites that can either promote or inhibit inflammation in the gut. By understanding how specific bacteria interact with the immune system and intestinal cells, researchers can potentially develop targeted therapies to modulate the microbiome and alleviate symptoms of gastrointestinal diseases. This growing body of evidence underscores the importance of maintaining a healthy gut microbiome for overall gastrointestinal health.

Probiotics and Gut Health

The influence of probiotics on gut health is a topic of great interest in the field of microbiology and human health. Probiotics, which are live microorganisms that confer health benefits when consumed in adequate amounts, have been shown to play a crucial role in maintaining the delicate balance of bacteria in the gut. By promoting the growth of beneficial bacteria and inhibiting the growth of harmful ones, probiotics can help support a healthy gut microbiome. Studies have demonstrated that probiotics can improve digestion, enhance immune function, and even reduce inflammation in the gut. Furthermore, research has suggested that probiotics may play a role in the prevention and management of certain gastrointestinal disorders, such as irritable bowel syndrome and inflammatory bowel disease. As our understanding of the human microbiome continues to grow, the potential for probiotics to positively impact gut health remains a promising area of research.

VI. MICROBIOME AND IMMUNE SYSTEM INTERACTION

The intricate relationship between the human microbiome and the immune system is a subject of growing interest and importance in the field of medical research. Recent studies have shed light on the ways in which the trillions of bacteria residing in our bodies interact with our immune cells, influencing our susceptibility to diseases and overall well-being. The microbiome not only helps to educate and train our immune system from an early age but also plays a crucial role in regulating its responses to pathogens. This dynamic interaction between the microbiome and the immune system is a delicate balance that, when disrupted, can lead to dysregulation and increase the risk of various autoimmune disorders. Understanding the intricate crosstalk between the microbiome and the immune system opens up new avenues for targeted therapies and interventions to manipulate this relationship for improved health outcomes. As research in this field continues to advance, the potential for harnessing the power of the microbiome to modulate immune responses and treat various diseases holds promise for the future of personalized medicine and preventive healthcare strategies.

Development of Immune System

As the human microbiome continues to garner increasing attention in the realm of healthcare and research, understanding the intricate development of the immune system is essential. The immune system, comprising a complex network of cells, tissues, and organs, plays a fundamental role in protecting the body against pathogens and maintaining homeostasis. During early life, the immune system undergoes a series of maturation processes, influenced by various factors including genetics, environmental exposures, and interactions with microbial communities. The establishment of a diverse and balanced microbiome in infancy is crucial for the proper development of the immune system, shaping its ability to distinguish between beneficial and harmful microorganisms. Dysbiosis, or microbial imbalance, has been linked to immune dysregulation and increased susceptibility to various diseases. Therefore, a deeper comprehension of how the human microbiome influences immune development is paramount in advancing our knowledge of health and disease.

Microbiome's Role in Immune Modulation

Research has shown that the human microbiome plays a crucial role in immune modulation. The microbiome, comprised of trillions of bacteria residing in our gut, skin, and other areas of the body, interacts with our immune system in complex ways. Through signaling molecules and metabolites produced by these bacteria, they can communicate with immune cells and influence their response to pathogens. For example, certain beneficial bacteria have been found to enhance the production of anti-inflammatory molecules, helping to regulate immune responses and prevent chronic inflammation. On the other hand, dysbiosis, an imbalance in the microbiome composition, has been linked to various immune-related disorders such as autoimmune diseases. By understanding the interactions between the microbiome and the immune system, researchers can develop new strategies for promoting health and treating immune-related conditions. Thus, the microbiome's role in immune modulation highlights the importance of maintaining a diverse and balanced microbial community for overall well-being.

Impact on Autoimmune Diseases

The impact of the human microbiome on autoimmune diseases is a complex and dynamic area of study that continues to intrigue researchers and medical professionals alike. The intricate interplay between the gut microbiota and the immune system is now being recognized as a key factor in the development and progression of autoimmune conditions. Studies have shown that dysbiosis, or an imbalance in the gut microbiome, can lead to increased inflammation and a breakdown of immune tolerance, which are hallmark features of autoimmune diseases. Furthermore, specific bacterial strains have been implicated in either promoting or protecting against autoimmune disorders, highlighting the potential for targeted interventions through probiotics or other microbiome-modulating therapies. Understanding the intimate relationship between the microbiome and autoimmune diseases holds great promise for novel diagnostic tools and personalized treatment strategies that could revolutionize the management of these complex conditions.

VII. THE SKIN MICROBIOME

The skin microbiome, known as the collection of microorganisms that inhabit the skin, has gained increasing attention in recent years for its pivotal role in maintaining skin health and function. Comprising a diverse array of bacteria, fungi, and viruses, the skin microbiome serves as a crucial barrier against pathogens and helps to regulate inflammation and immune responses. Research has shown that disruptions in the balance of the skin microbiome, known as dysbiosis, can lead to various skin disorders such as acne, eczema, and psoriasis. Understanding the intricacies of the skin microbiome not only sheds light on the complex interactions between microbes and the skin but also opens up new possibilities for therapeutic interventions to treat skin conditions. By elucidating the dynamics of the skin microbiome, we can harness its potential to improve skin health and overall well-being.

Composition and Function

The composition and function of the human microbiome are intricately intertwined, with the diverse array of bacteria colonizing various parts of our bodies playing vital roles in maintaining our health. These microbial communities are not just passive inhabitants; they actively participate in crucial physiological processes, such as nutrient metabolism, immune system modulation, and even influencing our mental health. The gut microbiota, for example, plays a pivotal role in the digestion and absorption of nutrients, as well as in the synthesis of essential vitamins. In addition, these microorganisms interact with our immune system, helping to train it and distinguish between harmful pathogens and beneficial commensals. Furthermore, recent research has highlighted the significant impact of the microbiome on mental health, with gut-brain communication pathways influencing mood and behavior. Understanding the composition and function of these bacterial communities is essential for developing targeted interventions to maintain a balanced microbiome and promote overall health and well-being.

Relationship with Skin Conditions

The relationship between the human microbiome and skin conditions is a complex and dynamic one. Numerous studies have shown that the composition of bacteria in the skin microbiome can influence the development and severity of various skin conditions, such as eczema, acne, and psoriasis. For example, dysbiosis in the skin microbiome, characterized by an imbalance of harmful and beneficial bacteria, has been linked to the pathogenesis of eczema. Certain types of bacteria, such as Staphylococcus aureus, have been found to exacerbate inflammation in eczema-prone individuals. In contrast, other beneficial bacteria, like Propionibacterium acnes, can contribute to maintaining a healthy skin barrier and reducing the risk of acne development. Understanding the intricate interplay between the microbiome and skin conditions holds great potential for developing novel therapeutic strategies that target the microbiome to improve skin health.

Therapeutic Approaches for Skin Health

Recent advancements in the field of dermatology have brought forth a variety of therapeutic approaches for enhancing skin health. One promising method is the use of probiotics, which are live bacteria and yeasts that are beneficial for the skin's microbiome. By introducing these friendly microorganisms, probiotics can help restore balance to the skin's ecosystem, promoting a healthy barrier function and reducing inflammation. Another innovative approach involves the use of prebiotics, which are compounds that serve as food for beneficial skin bacteria. By nourishing these good bacteria, prebiotics can help maintain a diverse and resilient microbiome, thereby supporting skin health. Additionally, personalized skincare treatments tailored to an individual's unique microbiome profile are gaining traction, allowing for targeted and effective interventions. These therapeutic approaches not only offer exciting possibilities for improving skin conditions but also underscore the immense potential of harnessing the power of the human microbiome in promoting overall health and well-being.

VIII. THE ORAL MICROBIOME

As we delve deeper into the complexities of the human microbiome, it becomes evident that the oral microbiome, specifically, plays a significant role in our overall health. The oral cavity harbors a diverse array of bacteria that interact with each other and with the host, influencing not only oral health but also systemic health. Studies have shown that dysbiosis in the oral microbiome can lead to various oral diseases such as periodontal disease, dental caries, and oral cancers. Furthermore, emerging research suggests that the oral microbiome may have implications beyond the oral cavity, impacting conditions such as cardiovascular disease, diabetes, and even neurodegenerative disorders. Understanding the intricate relationship between the oral microbiome and overall health is crucial in developing targeted interventions to promote oral and systemic health. By harnessing the power of these oral microbes, we may unlock new avenues for improving human health and well-being.

Composition and Its Unique Environment
Understanding the composition of the human microbiome involves recognizing the symbiotic relationship between bacteria and their unique environment within the body. It is essential to appreciate that the microbiome is not a static entity but a dynamic ecosystem that constantly adapts to internal and external stimuli. The diverse range of microbial species residing in different areas of the body, such as the gut, skin, and oral cavity, interact with each other and with host cells to maintain a delicate balance. This balance is crucial for a variety of physiologi-

cal functions, including digestion, immune response, and nutrient absorption. The environment within each body site influences the composition and activity of the microbiome, shaping its impact on health and disease. Therefore, comprehending the intricate interplay between composition and environment is essential for unlocking the therapeutic potential of targeting the microbiome to improve human health.

Link to Oral Diseases
Recent studies have highlighted the intricate link between the human microbiome and various oral diseases. The oral cavity is home to a diverse array of bacteria, which can either hinder or promote oral health. A balanced microbiome in the mouth is essential for maintaining healthy gums and teeth, as certain bacteria have been found to be associated with the development of conditions such as cavities, gingivitis, and periodontal disease. Dysbiosis in the oral microbiome, characterized by an imbalance in bacterial populations, can lead to an overgrowth of harmful bacteria, ultimately contributing to the progression of oral diseases. Understanding the dynamics of the oral microbiome and its impact on oral health is crucial for developing targeted therapies that can help prevent and treat these conditions effectively. By unraveling the complex interactions between bacteria in the oral cavity, researchers can pave the way for innovative strategies to promote good oral health and overall well-being.

Preventive Measures and Oral Health
Recent research has highlighted the importance of preventive measures in maintaining optimal oral health by preserving the delicate balance of the microbiome within the oral cavity. The

use of probiotics, specifically beneficial strains of bacteria such as lactobacilli and bifidobacteria, has shown promise in promoting oral health by inhibiting the growth of pathogenic bacteria and reducing the risk of dental caries and periodontal diseases. Furthermore, maintaining good oral hygiene practices, such as regular brushing and flossing, can help prevent the overgrowth of harmful bacteria that may disrupt the microbiome equilibrium. By incorporating preventive measures into daily oral care routines, individuals can support the diversity and stability of the oral microbiome, ultimately leading to improved overall oral health outcomes. Emphasizing preventive measures not only benefits oral health but also highlights the interconnectedness of the human microbiome in maintaining overall health and well-being.

IX. THE RESPIRATORY MICROBIOME

The respiratory microbiome, representing the microbial communities in the lungs and airways, has garnered increasing attention for its potential implications on human health. Recent research has revealed that the respiratory microbiome is not only present in healthy individuals but also shapes the immune response and susceptibility to respiratory diseases. The diversity and composition of bacteria in the respiratory tract contribute to maintaining immune homeostasis, while dysbiosis in this niche has been linked to the development of respiratory conditions such as chronic obstructive pulmonary disease (COPD) and asthma. Understanding the dynamic interactions between the respiratory microbiome and the host may offer new avenues for therapeutic interventions to modulate immune responses and improve respiratory health. By further exploring the intricate relationship between the respiratory microbiome and disease pathogenesis, researchers can potentially harness the power of microbial communities to promote respiratory well-being.

Microbial Communities in the Respiratory Tract
Recent research has shed light on the complex microbial communities harbored within the human respiratory tract. These communities, composed of diverse bacteria, viruses, and fungi, play a significant role in maintaining the delicate balance of the respiratory system. The respiratory microbiome has been shown to interact with the host immune system, influencing susceptibility to infections and respiratory diseases. Furthermore, disruptions in the composition of these microbial communities have been linked to various respiratory conditions, including asthma,

chronic obstructive pulmonary disease (COPD), and cystic fibrosis. By understanding the intricate interplay between the respiratory microbiome and the host, researchers can potentially develop novel therapeutic strategies for respiratory diseases. This emerging field of study underscores the importance of investigating the impact of microbial communities on respiratory health and highlights the potential for microbiome-based interventions to improve patient outcomes.

Impact on Respiratory Health

Evidence suggests that the human microbiome exerts a significant impact on respiratory health. The intricate balance of beneficial bacteria in the respiratory tract plays a crucial role in protecting against pathogens and maintaining proper immune function. When this balance is disrupted, through factors such as antibiotic use or environmental exposures, it can lead to respiratory infections, allergies, and even chronic respiratory diseases. Studies have shown that alterations in the microbiome composition in the lungs can contribute to the development of conditions like asthma and chronic obstructive pulmonary disease (COPD). Understanding the relationship between the microbiome and respiratory health is essential for developing targeted interventions that can restore balance and promote respiratory well-being. By exploring the complexities of the human microbiome, researchers can uncover new strategies for preventing and managing respiratory conditions, ultimately improving health outcomes for individuals.

Future Directions in Respiratory Microbiome Research

As we look towards the future of respiratory microbiome research, it is imperative to consider the potential impact of advancements in technology and methodology. With the advent of high-throughput sequencing techniques, researchers are now able to study the respiratory microbiome in greater detail, identifying specific bacterial species and their roles in respiratory health and disease with higher precision. Furthermore, integrating multi-omics approaches, such as metagenomics, metatranscriptomics, and metabolomics, can provide a more comprehensive understanding of the complex interactions within the respiratory microbiome. Collaborative efforts between microbiologists, immunologists, pulmonologists, and bioinformaticians will be crucial in unraveling the intricate relationships between the respiratory microbiome and host immunity. By embracing interdisciplinary collaborations and innovative research strategies, future studies in respiratory microbiome research have the potential to revolutionize our understanding of respiratory diseases and pave the way for novel therapeutic interventions targeting the microbiome.

X. THE UROGENITAL MICROBIOME

The urogenital microbiome, comprising of microorganisms inhabiting the urinary and reproductive tracts, has gained increasing attention for its impact on human health. Recent research has revealed the intricate relationship between the urogenital microbiome and various health conditions, including urinary tract infections, sexually transmitted diseases, and infertility. Despite being initially thought to be sterile, these regions of the body are now recognized as complex ecosystems hosting diverse bacterial communities that interact with the host immune system and influence physiological processes. Dysbiosis in the urogenital microbiome has been linked to a range of disorders, highlighting the importance of maintaining a balanced microbial community in these areas. Understanding the composition and function of the urogenital microbiome is crucial for developing targeted therapies to promote health and prevent disease in these sensitive regions of the body.

Composition and Health Implications
Research has shown that the composition of the human microbiome has far-reaching implications for our health. The intricate balance of bacteria in our bodies not only aids in digestion and nutrient absorption but also plays a crucial role in immunity and disease prevention. Dysbiosis, or the disruption of this delicate balance, has been linked to a range of health issues, including inflammatory bowel disease, obesity, and even mental health disorders. By understanding how the composition of the microbiome influences our health, we can potentially uncover new ways to prevent and treat these conditions. Moreover, emerging

research suggests that manipulating the microbiome through probiotics, prebiotics, and fecal transplants may offer innovative therapeutic approaches for a variety of health concerns. As we continue to delve deeper into the complexity of the human microbiome, unlocking its potential may revolutionize healthcare practices and improve outcomes for countless individuals.

Influence on Reproductive Health

Emerging research has shed light on the significant influence of the human microbiome on reproductive health. The intricate interplay between the microbiota and the reproductive system has been shown to impact fertility, pregnancy outcomes, and even the development of conditions such as endometriosis and polycystic ovary syndrome (PCOS). Studies have demonstrated that the composition of the vaginal microbiota can influence the likelihood of successful implantation and the risk of preterm birth. Furthermore, disruptions in the gut microbiome have been linked to hormonal imbalances and inflammation, which can have profound effects on reproductive function. Understanding the role of the microbiome in reproductive health presents exciting opportunities for developing innovative diagnostic tools and interventions to optimize fertility outcomes and improve maternal and infant health. As we delve deeper into this complex relationship, harnessing the power of the microbiome may hold the key to addressing a myriad of reproductive health challenges.

Strategies for Managing Urogenital Health

Recent research has identified several strategies for managing urogenital health by manipulating the microbiome. One approach involves probiotic supplementation, which aims to restore healthy bacterial populations in the urogenital tract. Studies have shown that certain strains of lactobacilli can inhibit the growth of pathogenic bacteria, thereby reducing the risk of infections such as urinary tract infections and bacterial vaginosis. Another strategy is the use of prebiotics, which are non-digestible fibers that promote the growth of beneficial bacteria in the gut and urogenital tract. By providing a favorable environment for good bacteria to thrive, prebiotics can help maintain a healthy microbial balance and prevent the overgrowth of harmful species. Additionally, personalized microbiome-based therapies are being developed to target specific imbalances in the urogenital microbiota, offering a tailored approach to managing urogenital health. These innovative strategies hold promise for improving outcomes in urogenital health and preventing associated complications.

XI. MICROBIOME AND METABOLIC SYNDROME

Emerging research suggests a strong connection between the human microbiome and metabolic syndrome, a cluster of conditions that increase the risk of heart disease, stroke, and diabetes. The microbiome, consisting of trillions of bacteria residing in the gut, has been found to play a pivotal role in regulating metabolism and inflammation. Dysbiosis, an imbalance in the gut flora, has been linked to the development of metabolic syndrome, as certain bacteria can promote inflammation and insulin resistance. Probiotics and prebiotics have shown promise in rebalancing the microbiome and improving metabolic health. Additionally, dietary interventions such as fiber-rich foods and fermented products can positively impact gut bacteria composition and metabolic functions. Understanding the intricate dance between the microbiome and metabolic syndrome opens up new possibilities for therapeutic interventions and preventive strategies in managing this complex condition.

Influence on Obesity and Diabetes

Recent research has shown a significant correlation between the composition of the human microbiome and the prevalence of obesity and diabetes. Studies have demonstrated that individuals with a higher ratio of Firmicutes to Bacteroidetes in their gut microbiota are more likely to be obese. These bacteria play a role in extracting energy from food and storing it as fat. Additionally, specific strains of gut bacteria have been linked to insulin resistance and inflammation, two key factors in the development of diabetes. The microbiome can influence metabolic processes and immune responses, ultimately impacting susceptibility to these chronic diseases. Understanding the intricate relationship between the microbiome and obesity and diabetes offers new avenues for therapeutic interventions, such as probiotics and dietary interventions aimed at modulating the gut microbiota to improve metabolic health. By targeting the microbiome, personalized treatment strategies could potentially help mitigate the growing burden of obesity and diabetes in modern societies.

Mechanisms Linking Microbiome to Metabolism

While the specific mechanisms linking the microbiome to metabolism are still being elucidated, several key pathways have been identified. One crucial interaction is the production of short-chain fatty acids (SCFAs) by gut microbiota, which can directly influence host metabolism by serving as an energy source for colonocytes and affecting lipid and glucose metabolism. Additionally, the gut microbiome can modulate bile acid metabolism, which plays a critical role in lipid digestion and absorption. Furthermore, certain gut bacteria have been shown to impact the production of neurotransmitters like serotonin and gamma-aminobutyric acid (GABA), which can influence appetite and food intake. These intricate interactions highlight the complexity of the microbiome-metabolism axis and underscore the importance of further research in this field to better understand how manipulating the microbiome can potentially be utilized to improve metabolic health.

Intervention Strategies

In considering intervention strategies for shaping the human microbiome, it is imperative to acknowledge the complexity and diversity of microbial communities residing within the body. As such, a personalized approach tailored to an individual's unique microbial composition may prove most effective. Utilizing targeted probiotics or prebiotics that support the growth of beneficial bacteria while inhibiting harmful pathogens could be a promising intervention strategy. Furthermore, dietary modifications that promote a diverse and healthy microbiome, such as increasing fiber intake or consuming fermented foods, may offer long-term benefits for overall health. Implementing lifestyle changes, such as stress reduction techniques or regular exercise, can also positively impact the microbiome. In conjunction with traditional medical treatments, these intervention strategies have the potential to not only optimize microbial balance but also enhance immune function and mitigate various health conditions associated with dysbiosis.

XII. MICROBIOME AND CARDIOVASCULAR HEALTH

The influence of the microbiome on cardiovascular health has gained increasing attention in recent years. Research indicates that disturbances in the composition of gut bacteria can lead to systemic inflammation, a key contributor to the development of cardiovascular diseases. The intricate interaction between the microbiome and the immune system plays a vital role in this process, with certain bacterial species either promoting or mitigating inflammation. For example, an abundance of beneficial bacteria like Lactobacillus and Bifidobacterium has been associated with a reduced risk of cardiovascular events. On the other hand, an overgrowth of harmful bacteria such as Prevotella and Desulfovibrio can contribute to inflammation and atherosclerosis. Understanding the complex relationship between the microbiome and cardiovascular health offers exciting possibilities for targeted interventions and personalized treatments, paving the way for innovative approaches to managing and preventing cardiovascular diseases.

Impact on Heart Disease
The intricate relationship between the human microbiome and heart disease has garnered significant attention in recent years. Research suggests that the composition of gut bacteria can influence various risk factors for heart disease, such as obesity, inflammation, and cholesterol levels. By producing metabolites that interact with our immune system and modulate inflammation, gut bacteria play a crucial role in the development and progression of cardiovascular conditions. Additionally, certain

strains of bacteria have been linked to atherosclerosis, a major contributor to heart disease. Understanding the impact of the microbiome on heart health opens up new possibilities for therapeutic interventions focused on promoting a balanced and diverse microbial community. By harnessing the potential of the microbiome, we may be able to develop innovative strategies for preventing and managing heart disease, ultimately improving the overall health and well-being of individuals.

Mechanisms of Influence

As we explore the mechanisms of influence that bacteria within the human microbiome have on our health, it becomes apparent that their interactions extend far beyond just the gut. These microorganisms produce a plethora of metabolites that can directly impact various physiological processes throughout the body, from immune function to neurological activity. Through the production of short-chain fatty acids, for example, gut bacteria can modulate the immune response, influencing inflammation levels and contributing to the maintenance of overall health. Moreover, certain strains of bacteria have been found to produce neurotransmitters and other neuroactive compounds, potentially affecting mood and cognitive function. By understanding the complex web of interactions between our microbiome and various bodily systems, we can harness the potential of these bacteria to promote health and prevent disease. The intricate mechanisms by which our microbiome exerts its influence underscore the importance of further research in uncovering the full extent of their impact on human health.

Potential for Therapeutic Interventions

In conclusion, the potential for therapeutic interventions targeting the human microbiome holds great promise for the future of medicine. By understanding the intricate relationship between the microbiota and our health, researchers and healthcare professionals can develop innovative treatments that target specific bacterial populations to restore balance and prevent diseases. These interventions can range from probiotics and prebiotics to fecal microbiota transplantation and precision microbiome editing. The diversity and complexity of the human microbiome require tailored and personalized approaches to address individual health needs effectively. As we continue to unravel the mysteries of the microbiome, the development of targeted therapeutic interventions will revolutionize the way we approach preventive and curative medicine, offering new hope for patients suffering from a wide range of conditions. Ultimately, harnessing the power of the human microbiome has the potential to transform healthcare and improve outcomes for millions worldwide.

XIII. MICROBIOME AND MENTAL HEALTH

Emerging research has provided compelling evidence of the intricate connection between the microbiome and mental health. The human gut microbiota, in particular, has been implicated in various mental health conditions such as anxiety, depression, and even neurodegenerative disorders. The bidirectional communication between the gut and the brain, known as the gut-brain axis, highlights the crucial role of gut microbiota in modulating neurological functions and behavior. Through the production of neurotransmitters, such as serotonin and dopamine, gut bacteria influence mood, cognition, and stress response. Dysbiosis, an imbalance in the gut microbiota composition, has been associated with mental health disorders, further underscoring the importance of maintaining a diverse and healthy microbial community. Understanding the intricate interplay between the microbiome and mental health opens new avenues for therapeutic interventions, emphasizing the potential of targeting gut microbiota for the treatment and prevention of mental health disorders.

Concept of the Gut-Brain Axis
In recent years, the concept of the gut-brain axis has gained significant attention in the field of microbiome research. This intricate bidirectional communication system between the gut and the brain involves a complex network of neurons, immune cells, hormones, and neurotransmitters. The gut microbiota, composed of trillions of bacteria residing in the gastrointestinal tract, plays a crucial role in modulating this axis. These bacteria produce various metabolites and signaling molecules that can

influence neural, immune, and endocrine pathways, ultimately impacting brain function and behavior. Studies have shown that disruptions in the gut microbiota composition, known as dysbiosis, may contribute to the development of neurological disorders, mood disorders, and cognitive impairments. Understanding the mechanisms underlying the gut-brain axis can provide valuable insights into potential therapeutic strategies for treating a wide range of conditions that extend beyond the gut.

Microbiome's Role in Mental Disorders

The intricate interplay between the human microbiome and mental health has been increasingly recognized in recent years. Research has revealed a bidirectional communication system between the gut microbiota and the brain, known as the gut-brain axis, which influences various physiological processes including mood, cognition, and behavior. Dysbiosis, or imbalance in gut bacteria composition, has been linked to the development of mental disorders such as depression, anxiety, and even neurodegenerative diseases. The microbiome is thought to impact mental health through several mechanisms, including the production of neurotransmitters, modulation of the immune system, and regulation of inflammation. Understanding the role of the microbiome in mental disorders offers new avenues for therapeutic interventions, such as probiotics, prebiotics, and dietary modifications, that target the gut-brain axis to promote mental well-being. Further studies exploring this complex relationship are essential for advancing our understanding and treatment of mental health conditions.

Probiotics and Mental Health Treatments

Emerging research has shed light on the potential of probiotics as a novel approach to mental health treatments. Probiotics, once primarily associated with gut health, are now being recognized for their impact on the gut-brain axis and the communication between the microbiome and the brain. Studies have shown that certain strains of beneficial bacteria can exert positive effects on mood, anxiety, and even cognitive function. By modulating the gut microbiota, probiotics may influence the production of neurotransmitters such as serotonin and gamma-aminobutyric acid (GABA), which play key roles in regulating mood and stress responses. Moreover, probiotics have been investigated for their anti-inflammatory properties, as gut dysbiosis and inflammation have been linked to various mental health disorders. Incorporating probiotics into mental health treatments could offer a promising avenue for improving symptoms and enhancing overall well-being. Further research is warranted to elucidate the specific mechanisms by which probiotics exert their effects on mental health and to optimize their therapeutic potential.

XIV. MICROBIOME AND CANCER

Recent studies have shed light on the intricate relationship between the human microbiome and cancer. The microbiome, consisting of trillions of bacteria residing within the body, has been found to play a significant role in the development and progression of various types of cancer. Dysbiosis, an imbalance in the microbial community, has been linked to increased inflammation and immune dysregulation, both of which are known to contribute to carcinogenesis. Moreover, certain bacteria have been identified as either promoting or suppressing tumor growth through mechanisms such as altering the microenvironment, influencing host immune responses, or producing metabolites that impact cancer cell behavior. Understanding the complex interactions between the microbiome and cancer may pave the way for novel therapeutic strategies, such as targeting specific bacteria or modulating the microbiome to enhance treatment efficacy and improve patient outcomes. As research in this field continues to advance, harnessing the potential of the microbiome in cancer management holds great promise for personalized medicine and improved patient care.

Influence on Cancer Development
Emerging research has highlighted the significant influence of the human microbiome on the development and progression of cancer. The intricate interplay between the microbiota and the host immune system can either promote or inhibit tumorigenesis. Dysbiosis, characterized by an imbalance in the microbial ecosystem, has been linked to chronic inflammation, which is a key player in carcinogenesis. Specific bacterial species have

been identified as potential culprits in the pathogenesis of various cancers by producing genotoxins, promoting cell proliferation, or modulating the immune response to favor tumor growth. Conversely, certain beneficial bacteria can enhance anti-tumor immune responses and produce metabolites that inhibit tumor cell growth. Understanding the complex relationship between the microbiome and cancer development is crucial for developing novel therapeutic strategies that manipulate the microbiota to prevent or treat cancer effectively. Further research in this field holds great promise for revolutionizing cancer treatment and improving patient outcomes.

Microbiome as a Diagnostic Tool

Recent advancements in the field of microbiome research have shed light on the potential of utilizing the human microbiome as a diagnostic tool. By analyzing the composition and diversity of bacteria residing in various parts of the body, researchers can identify specific patterns or dysbiosis that may be indicative of certain diseases or conditions. For example, alterations in the gut microbiota have been associated with inflammatory bowel diseases, obesity, and even neurological disorders. Through the use of high-throughput sequencing techniques and sophisticated bioinformatics tools, scientists can now profile the microbiome with unprecedented accuracy and resolution. This wealth of information holds promise for developing personalized diagnostic methods that are not only non-invasive but also highly sensitive and specific. As we continue to unravel the intricate relationship between the microbiome and human health, the potential for using bacteria as biomarkers for disease detection and monitoring becomes increasingly compelling.

Microbiome-targeted Therapies

Emerging research in the field of human microbiome has highlighted the potential of microbiome-targeted therapies to revolutionize healthcare. By focusing on manipulating the composition of the microbiota, these therapies aim to restore balance and promote health. One promising approach involves the use of prebiotics, probiotics, and synbiotics to enhance the growth of beneficial bacteria and suppress pathogenic microbes in the gut. These interventions can help improve gastrointestinal health, modulate immune responses, and even impact mental well-being. Furthermore, advancements in precision medicine have enabled the development of personalized microbiome-based treatments tailored to an individual's unique microbial profile. By harnessing the power of the microbiome, researchers are paving the way for innovative therapeutic strategies that have the potential to transform the landscape of modern medicine. As we continue to unravel the complexities of the human microbiome, the prospect of microbiome-targeted therapies offers new avenues for enhancing human health and combating a range of diseases.

XV. MICROBIOME AND AGING

Recent scientific research has shed light on the intricate relationship between the human microbiome and the process of aging. As individuals age, changes occur in the composition and diversity of the microbiota residing in our bodies, which can have significant implications for health outcomes. The microbiome's role in modulating inflammation, nutrient absorption, metabolism, and immune function becomes increasingly crucial as we age. With advancing age, there is a decline in microbial diversity and stability, leading to an imbalance in the gut ecosystem known as dysbiosis. This dysbiotic state has been linked to various age-related diseases, including cardiovascular disease, cognitive decline, and frailty. Understanding the dynamics of the microbiome in the context of aging holds promise for developing targeted interventions to promote healthy aging and improve overall well-being. By elucidating the mechanisms through which the microbiome influences the aging process, we can potentially harness the therapeutic potential of microbiota-targeted interventions to enhance healthspan and longevity.

Changes in Microbiome Over Life Span

The human microbiome undergoes dynamic changes over the course of a lifespan, influenced by various factors such as diet, environment, genetics, and aging. In early life, the microbiome is established through interactions with the mother during birth and breastfeeding, shaping the composition and diversity of microbial communities in the gut. As individuals age, the microbiome continues to evolve, with fluctuations in bacterial composition and function. These changes may impact health outcomes, as alterations in the microbiome have been linked to various diseases and conditions, including obesity, autoimmune disorders, and metabolic disorders. Understanding the trajectory of microbiome changes over the lifespan is essential for developing personalized approaches to maintaining a healthy microbiome and promoting overall well-being. By elucidating these patterns, researchers can uncover strategies to support microbial diversity and balance at different life stages, potentially mitigating the risk of disease and optimizing health outcomes.

Impact on Age-related Diseases

Emerging research suggests that the human microbiome may have a significant impact on age-related diseases. As individuals age, changes in the composition of the microbiome can lead to dysbiosis, which is associated with various health issues. The gut microbiota, in particular, plays a crucial role in regulating inflammation, metabolic function, and immune responses, all of which are key factors in the development of age-related diseases such as diabetes, cardiovascular disease, and neurodegenerative disorders. By maintaining a healthy balance of beneficial bacteria in the gut, it may be possible to mitigate the risk of these conditions and promote healthy aging. Understanding the intricate relationship between the microbiome and age-related diseases could pave the way for innovative therapeutic interventions that target the microbiota to prevent or treat these conditions effectively. This highlights the importance of further research in this field to unravel the full extent of the microbiome's impact on aging and disease.

Potential Interventions to Promote Healthy Aging

In considering potential interventions to promote healthy aging, one promising avenue lies in the manipulation of the gut microbiome. The composition and diversity of gut bacteria have been linked to a range of age-related conditions, from cognitive decline to inflammation and metabolic disorders. By modulating the gut microbiota through dietary interventions, such as the consumption of prebiotics, probiotics, and fermented foods, it may be possible to promote a healthier balance of gut bacteria and reduce the risk of age-related diseases. In addition to diet, factors like physical activity, stress management, and sleep patterns can also influence the gut microbiome and contribute to healthy aging. These interventions offer a holistic approach to promoting well-being in older adults by addressing the intricate relationship between gut health and overall physiological function. As our understanding of the human microbiome continues to evolve, targeted interventions to support healthy aging through gut health may hold promise for improving quality of life in aging populations.

XVI. PEDIATRIC MICROBIOME AND DEVELOPMENT

The development of the pediatric microbiome is a critical aspect of early childhood health and well-being. During the first years of life, the gut microbiome undergoes significant changes, influenced by factors such as diet, genetics, and environmental exposures. These early microbial communities play a crucial role in shaping the immune system and metabolic processes, impacting long-term health outcomes. Research has shown that disruptions in the pediatric microbiome can lead to various health issues, including allergies, obesity, and autoimmune diseases. Understanding the intricacies of how the microbiome evolves in early childhood is essential for developing interventions to promote optimal health and prevent disease later in life. By exploring the dynamic relationship between the pediatric microbiome and development, we can uncover new insights into the complex interplay between bacteria in the body and human health. This knowledge can ultimately lead to personalized strategies for supporting microbiome health and improving overall well-being from infancy through adulthood.

Establishment of the Microbiome in Infancy
As infants transition from the sterile environment of the womb to the outside world, the colonization of their microbiome begins in earnest. During birth, a baby's first exposure to microbes occurs through the mother's birth canal or during a cesarean section, influencing the initial composition of their gut microbiota. Subsequently, feeding practices, whether breast milk or formula, further shape the microbiome by providing specific nutrients

that promote the growth of beneficial bacterial species. The establishment of a diverse and resilient microbiome in infancy is crucial for the development of immune tolerance, metabolic functions, and overall health later in life. Notably, disruptions in the early colonization of the microbiome have been linked to various health conditions, highlighting the importance of understanding and supporting the establishment of a healthy microbiome from infancy.

Impact on Child Development and Health

One significant area where the human microbiome exerts a pronounced impact is on child development and health. During early life, the colonization of the gut by various bacterial species plays a crucial role in shaping the immune system and metabolic processes. Disruptions in this process, whether due to factors such as cesarean section delivery, formula feeding, or antibiotic use, can have long-lasting effects on a child's health. Research has shown that alterations in the composition of the gut microbiota in infancy have been associated with an increased risk of various health conditions, including allergies, asthma, obesity, and even neurological disorders. Furthermore, the interplay between the gut microbiome and the developing brain has been a topic of growing interest, with studies suggesting that disruptions in the gut-brain axis may contribute to mental health issues in children. Therefore, understanding the role of the human microbiome in child development is essential for promoting optimal health outcomes in the early years of life.

Strategies for Optimizing Pediatric Health

Emerging research suggests that optimizing pediatric health involves a multifaceted approach that goes beyond traditional medical interventions. One effective strategy is to focus on promoting a healthy microbiome in children. This can be done through various means, such as encouraging breastfeeding, which provides essential nutrients for beneficial gut bacteria to thrive. Additionally, avoiding unnecessary antibiotics in early childhood can help preserve the delicate balance of microbes in the body. Furthermore, incorporating a diverse range of fruits, vegetables, and whole grains into a child's diet can support a healthy microbiome. Physical activity and outdoor play have also been shown to positively impact gut health. By implementing these strategies, caregivers and healthcare providers can support the development of a robust microbiome in children, which is crucial for their overall health and well-being.

XVII. DIET AND THE MICROBIOME

Recent research has shed light on the intricate relationship between diet and the microbiome, highlighting the significant impact of food choices on the composition and function of our gut bacteria. Specifically, a diet rich in fiber has been linked to a diverse and healthy microbiome, promoting the growth of beneficial gut bacteria that can help regulate inflammation and improve overall digestive health. Conversely, a diet high in saturated fats and sugars has been shown to alter the microbiome in a way that promotes inflammation and contributes to chronic diseases such as obesity and diabetes. By understanding how different types of food affect the microbiome, we can make informed dietary choices that support a healthier balance of gut bacteria and potentially reduce the risk of various health conditions. This highlights the importance of considering diet as a key factor in maintaining a thriving microbiome and ultimately, optimizing our overall health and well-being.

Effects of Dietary Choices

The effects of dietary choices on the human microbiome are profound and far-reaching. Research has shown that the foods we consume on a daily basis have a direct impact on the composition and diversity of our gut bacteria. A diet high in processed foods, sugar, and saturated fats has been linked to an imbalance in the gut microbiota, which can lead to a host of health issues such as obesity, diabetes, and inflammatory bowel diseases. On the other hand, a diet rich in fruits, vegetables, whole grains, and lean proteins promotes the growth of beneficial bacteria in the gut, which in turn can improve digestion, boost immunity, and reduce the risk of chronic diseases. Therefore, making informed dietary choices is essential for maintaining a healthy microbiome and overall well-being. By paying attention to what we eat, we can positively influence the delicate balance of bacteria in our bodies and support optimal health.

Diet-based Modulation of the Microbiome

Numerous studies have shown that diet plays a crucial role in shaping the composition and function of the gut microbiome. Through a process called diet-based modulation, specific dietary components can promote the growth of beneficial bacteria while inhibiting the proliferation of harmful microbes. For instance, a diet rich in fiber has been linked to an increase in the abundance of beneficial bacteria such as Bifidobacteria and Lactobacilli, which are known for their beneficial effects on host health. Conversely, diets high in saturated fats and sugars have been associated with a decrease in bacterial diversity and an overgrowth of potentially pathogenic species. These findings highlight the intricate interplay between diet and the microbiome, underscoring the importance of dietary interventions as a means of modulating microbial communities to promote overall health and prevent disease. By understanding how different dietary components impact the microbiome, personalized dietary strategies can be developed to optimize microbial diversity and function, ultimately leading to improved health outcomes.

Recommendations for Microbiome-friendly Diets

It is essential to consider various recommendations for incorporating microbiome-friendly diets into everyday life. Firstly, individuals should prioritize a diverse and balanced intake of fiber-rich foods, as these are known to support a healthy gut microbiome. Foods such as fruits, vegetables, whole grains, and legumes can promote the growth of beneficial bacteria in the gut. Additionally, incorporating fermented foods into the diet, such as yogurt, kefir, and sauerkraut, can introduce probiotics that contribute to a diverse microbiome. Moreover, reducing the consumption of processed foods and added sugars is crucial, as these can negatively impact the diversity and composition of the gut microbiome. Overall, a diet rich in whole, plant-based foods, combined with probiotic-rich options, can help maintain a healthy microbiome and support overall health and well-being. By following these recommendations, individuals can cultivate a microbiome-friendly diet that promotes optimal microbial balance and function within the body.

XVIII. ANTIBIOTICS AND THE MICROBIOME

It is well-established that antibiotics play a vital role in combating bacterial infections and saving lives. However, their use can have significant consequences on the delicate balance of the human microbiome. The microbiome, consisting of trillions of bacteria that reside in our bodies, plays a crucial role in maintaining our health by aiding in digestion, regulating the immune system, and even influencing mental health. When antibiotics are introduced into the system, they do not discriminate between harmful bacteria causing infection and beneficial bacteria essential for our well-being. This indiscriminate destruction can lead to disruptions in the microbiome, resulting in long-lasting effects on overall health. Moreover, the overuse of antibiotics has been linked to the rise of antibiotic-resistant bacteria, posing a serious threat to public health. Therefore, it is imperative for healthcare providers to consider the impact of antibiotics on the microbiome and to use them judiciously to preserve this intricate ecosystem within us.

Impact of Antibiotic Use
One of the key impacts of antibiotic use on the human microbiome is the disruption it causes to the delicate balance of microbial communities in the body. Antibiotics are designed to target and destroy harmful bacteria, but in the process, they also indiscriminately kill off beneficial bacteria that are essential for maintaining our health. This disruption can lead to a condition known as dysbiosis, where harmful bacteria can overgrow and

cause a range of health problems. Additionally, repeated or prolonged antibiotic use can result in antibiotic resistance, making it harder to treat bacterial infections in the future. As our understanding of the human microbiome deepens, it becomes increasingly clear that preserving the diversity and balance of our microbial populations is crucial for our overall well-being. Therefore, it is important to use antibiotics judiciously and explore alternative strategies to protect and support the health of our microbiome.

Strategies to Mitigate Negative Effects

Considering the significant impact of the human microbiome on our health, it is crucial to explore effective strategies to mitigate any negative effects resulting from imbalances in bacterial populations. One such strategy involves probiotic supplementation, which introduces beneficial bacteria into the gut to restore microbial equilibrium. By promoting the growth of these helpful microbes, probiotics can help combat harmful pathogens and reduce inflammation in the body. Additionally, diet plays a key role in shaping the composition of the microbiome, so dietary interventions such as increasing fiber intake and consuming fermented foods can support a diverse and healthy bacterial community. Another important approach is the use of prebiotics, which provide the necessary nourishment for beneficial bacteria to thrive. By implementing these strategies, we can actively work towards maintaining a harmonious microbiome and safeguarding our overall health and well-being.

Future of Antibiotic Policies

The future of antibiotic policies will be crucial in preserving the efficacy of these essential medications in combating bacterial infections. As antibiotic resistance continues to rise, with some strains becoming resistant to multiple antibiotics, there is a pressing need for a global collaborative effort to address this public health threat. One approach could involve implementing stricter regulations on antibiotic use in agriculture to reduce the spread of resistant bacteria through food sources. In healthcare settings, promoting antibiotic stewardship programs that encourage judicious use of these medications can help prevent the emergence of resistance. Additionally, investing in research and development of new antibiotic agents and alternative treatments, such as phage therapy or immunotherapy, will be essential to stay ahead of the evolving bacterial resistance mechanisms. Ultimately, a comprehensive and multifaceted approach to antibiotic policies is imperative to ensure the continued effectiveness of these life-saving medications.

XIX. FECAL MICROBIOTA TRANSPLANTATION (FMT)

Recent advancements in the field of microbiology have shed light on the potential benefits of Fecal Microbiota Transplantation (FMT) in treating various gastrointestinal disorders. FMT involves the transfer of fecal material from a healthy donor to a recipient in order to restore a healthy balance of gut bacteria. This procedure has shown promising results in the treatment of conditions such as Clostridium difficile infection, inflammatory bowel disease, and irritable bowel syndrome. By introducing a diverse range of beneficial microbes into the recipient's gut, FMT can help rebalance the microbiota and promote a healthy gut environment. While FMT is still considered a novel treatment approach, ongoing research is uncovering its potential applications beyond gastrointestinal disorders, including its role in immune system modulation and metabolic health. As our understanding of the human microbiome continues to deepen, FMT holds considerable promise as a therapeutic intervention with far-reaching implications for human health and well-being.

Principles and Procedures

The Principles and Procedures governing research on the human microbiome are essential in advancing our understanding of how bacteria impact our health. By adhering to rigorous scientific methodologies and ethical guidelines, researchers can uncover the intricate relationships between microbial communities and the human body. Through the application of cutting-edge technologies such as metagenomics and bioinformatics, scientists can decode the genetic makeup of these microorganisms and identify their functions within the host. These approaches provide invaluable insights into the complex interplay between microbiota and host cells, shedding light on mechanisms underlying various diseases and health conditions. Additionally, standardized protocols for sample collection, processing, and analysis ensure the reproducibility of results, allowing for robust comparisons across studies. By upholding these principles and procedures, researchers can pave the way for personalized interventions and therapies that harness the power of the human microbiome to improve health outcomes.

Clinical Applications

In the realm of clinical applications, the study of the human microbiome offers a wealth of potential for advancing medical treatment and personalized healthcare. Researchers are exploring how the microbiome can influence drug metabolism, immune system function, and even mental health. By understanding the intricate interplay between the body's microbial inhabitants and various physiological processes, healthcare providers can develop targeted interventions to treat a wide range of conditions more effectively. For example, manipulating the microbiome through probiotics or fecal transplants has shown promise in treating certain gastrointestinal disorders. Furthermore, the ability to analyze an individual's microbiome profile may pave the way for personalized treatment strategies that take into account each person's unique microbial composition. As the field of microbiome research continues to evolve, the clinical implications are vast, offering new avenues for improving health outcomes and enhancing the quality of patient care.

Ethical and Regulatory Considerations

In considering the ethical and regulatory aspects of human microbiome research, it is crucial to acknowledge the complexity of this field. As scientists delve deeper into the interactions between microbial communities and human health, questions arise concerning the potential for unintended consequences and the need for responsible oversight. Ethical dilemmas may arise surrounding issues such as informed consent, privacy concerns, and the equitable distribution of benefits from microbiome research. Additionally, regulatory frameworks must be established to ensure that research is conducted in a manner that is transparent, ethical, and in line with established standards. It is imperative that researchers, policymakers, and stakeholders work together to address these considerations proactively, in order to ensure that advancements in the understanding of the human microbiome are achieved in an ethical and responsible manner, ultimately benefiting society as a whole.

XX. ETHICAL CONSIDERATIONS IN MICROBIOME RESEARCH

Ethical considerations in microbiome research are paramount due to the sensitive nature of human subjects involved. Researchers must navigate complex issues such as informed consent, privacy, and beneficence when conducting studies that involve the human microbiome. Informed consent is crucial to ensure that participants fully understand the risks and benefits of participating in research, especially when it comes to microbiome studies that involve intimate bodily samples. Privacy concerns arise as the microbiome data collected is inherently personal and can reveal information about an individual's health status or genetic predispositions. Moreover, researchers must uphold the principle of beneficence by ensuring that the research benefits outweigh any potential harms to participants. Striking a balance between advancing scientific knowledge and protecting the rights and well-being of individuals participating in microbiome research is a challenge that requires careful consideration and adherence to ethical guidelines.

Privacy and Data Management

As our understanding of the human microbiome deepens, concerns surrounding privacy and data management in this field become increasingly relevant and complex. The intricate relationship between the bacteria in our bodies and our health necessitates the collection and analysis of vast amounts of personal data, raising ethical questions about consent, ownership, and potential misuse. Privacy breaches and unauthorized access to sensitive information pose significant risks to individuals, highlighting the need for robust data protection measures and transparency in research practices. Striking a balance between advancing scientific knowledge through data sharing and respecting individuals' rights to privacy is a crucial challenge that must be navigated with care and integrity. Establishing clear guidelines and standards for the ethical collection, storage, and use of microbiome data is essential to ensure that research in this field upholds the highest ethical standards and benefits society as a whole.

Consent and Participation

The concept of consent and participation within the context of human microbiome research is crucial to ethical considerations and ensuring the autonomy of individuals involved. Informed consent is essential when conducting studies that involve the collection and analysis of biological samples from participants, as they must be fully aware of the potential risks, benefits, and purposes of the research. Participants should have the right to withdrawal at any time without consequences and have their privacy and confidentiality protected. Furthermore, active participation from individuals in research projects can lead to more accurate and meaningful results, as their cooperation and adherence to study protocols are essential for successful data collection. Researchers must prioritize transparency and communication with participants to foster a trusting and respectful relationship, ultimately contributing to the advancement of knowledge in the field of human microbiome research.

Implications of Microbiome Manipulation

The implications of manipulating the human microbiome are

vast and complex. By understanding how various bacteria interact within our bodies, researchers can potentially develop targeted interventions to treat a wide range of health conditions. For example, altering the composition of the gut microbiota through probiotics or fecal transplants has shown promise in treating conditions such as inflammatory bowel disease and irritable bowel syndrome. However, the manipulation of the microbiome also raises ethical concerns, as the long-term effects of altering the delicate balance of bacteria in the body are not yet fully understood. Additionally, there is a need for further research to explore the potential risks and benefits of microbiome manipulation, ensuring that these interventions are both effective and safe for patients. Ultimately, the implications of microbiome manipulation highlight the need for careful consideration and continued investigation in this rapidly evolving field of study.

XXI. PUBLIC AWARENESS AND EDUCATION ON THE MICROBIOME

An essential aspect of harnessing the potential of the human microbiome lies in the realm of public awareness and education. Given the significant impact that microbial communities have on our health, it is crucial to disseminate accurate information and promote understanding among the general population. By enhancing public knowledge about the microbiome, individuals can make informed decisions regarding their lifestyle choices, such as diet, hygiene practices, and antibiotic use, that directly affect the balance of beneficial and harmful bacteria in their bodies. Moreover, increased awareness can also lead to more informed conversations with healthcare providers, ultimately improving the quality of healthcare and personalized treatment plans. Therefore, investing in educational initiatives that aim to raise public awareness about the microbiome is paramount in empowering individuals to take proactive steps towards optimizing their health and well-being.

Current Public Knowledge Levels

Understanding the current public knowledge levels concerning the human microbiome is essential in highlighting the importance of bacteria in our bodies for overall health. Studies have shown that many individuals are unaware of the intricate relationship between the microbiome and various aspects of health, such as immune function, digestion, and mental well-being. This lack of awareness can lead to detrimental health outcomes, as people may not realize the significance of maintaining a healthy balance of bacteria in their bodies. Public education and awareness campaigns are vital in bridging this knowledge gap and empowering individuals to make informed decisions about their health. By increasing awareness and understanding of the human microbiome, we can encourage proactive measures to support a diverse and thriving bacterial community within our bodies, ultimately leading to improved health outcomes and quality of life.

Importance of Educating the Public

In the realm of healthcare and wellness, the importance of educating the public about the human microbiome cannot be overstated. Understanding the intricacies of how bacteria within our bodies influence our health is crucial for promoting better health outcomes and disease prevention. By disseminating accurate information about the human microbiome to the general populace, individuals can make informed decisions about their lifestyles, diets, and medical treatments. This knowledge empowers individuals to take proactive steps in maintaining a healthy balance of bacteria within their bodies, which in turn can lead to improved immune function, digestion, and overall well-being. Furthermore, public education about the human microbiome can help dispel misconceptions and myths surrounding bacteria, fostering a greater appreciation for the beneficial roles these microorganisms play in our bodies. Ultimately, by prioritizing public education on the human microbiome, we can pave the way for a healthier future for all.

Strategies for Effective Communication

To effectively communicate the importance of the human microbiome and its impact on our health, strategies must be carefully employed. One such strategy is the use of clear and concise language to convey complex scientific concepts in a way that is easily understandable to a broader audience. This involves avoiding jargon and technical terms that may alienate or confuse lay readers. Additionally, utilizing visual aids such as charts, graphs, and diagrams can help illustrate key points and enhance comprehension. Another vital strategy is to tailor the communication to the specific audience, adapting the language and tone to suit their level of understanding and interest. By utilizing these strategies, researchers and healthcare professionals can effectively convey the significance of the human microbiome in promoting overall health and well-being, fostering a greater understanding and appreciation for this essential aspect of human biology.

XXII. REGULATORY CHALLENGES IN MICROBIOME RESEARCH

Research in the field of human microbiome has gained significant attention in recent years due to its potential to revolutionize healthcare. However, this burgeoning area of study is fraught with regulatory challenges that impede progress. One major obstacle is the lack of standardized protocols for sample collection, analysis, and interpretation in microbiome research. Without established guidelines, researchers may struggle to replicate findings or compare results across studies, hindering the advancement of the field. Additionally, the ethical considerations surrounding the use of human microbiome data raise important questions about privacy, consent, and data sharing. Striking a balance between promoting research innovation and protecting participants' rights is a delicate task that requires careful navigation. Addressing these regulatory challenges is essential for ensuring the integrity and reliability of microbiome research, which holds tremendous promise for improving human health.

Overview of Regulatory Landscape

Recent advancements in scientific research have shed light on the intricate relationship between the human microbiome and our health. Within this complex ecosystem of microorganisms that reside within our bodies, there exists a delicate balance that can profoundly impact various aspects of our well-being. The regulatory landscape governing research and applications related to the human microbiome is multifaceted, encompassing ethical considerations, legal frameworks, and adherence to scientific standards. As the field evolves rapidly, there is a growing need for clear guidelines and oversight to ensure the responsible and ethical advancement of microbiome-based treatments and therapies. Regulatory bodies play a crucial role in establishing guidelines for research, clinical trials, and commercialization of microbiome-related products, aiming to safeguard public health and prevent potential ethical dilemmas. This evolving regulatory landscape underscores the importance of balancing scientific progress with ethical considerations to harness the full potential of the human microbiome in improving health outcomes.

Challenges in Standardizing Protocols

When it comes to the human microbiome, one of the key challenges that researchers face is the standardization of protocols for studying this complex ecosystem. With hundreds of different bacterial species residing in various parts of the body, each with its unique functions and interactions, developing uniform methodologies for sampling, analyzing, and interpreting data becomes daunting. Variations in sample collection techniques, DNA extraction methods, sequencing technologies, and data analysis tools can lead to inconsistencies in results and hinder the reproducibility of studies. Moreover, the dynamic nature of the microbiome, influenced by numerous factors such as diet, lifestyle, and medications, adds another layer of complexity to standardization efforts. Despite these challenges, establishing standardized protocols is essential for comparing data across studies, advancing our understanding of the microbiome's role in health and disease, and ultimately translating research findings into clinical applications. Thus, ongoing collaboration among researchers and continuous refinement of methodologies are crucial in overcoming these obstacles and moving the field forward.

Future Directions in Regulation

With advancements in technology and a deeper understanding of the human microbiome, future directions in regulation are becoming increasingly important. As researchers continue to uncover the intricate relationship between the microbiome and human health, there is a growing need for regulatory frameworks to ensure the safe and effective use of microbiome-based therapies. Regulation in this field must balance the need for innovation and scientific progress with the importance of patient safety and ethical considerations. As we move forward, it will be crucial to establish standardized guidelines for the development, testing, and implementation of microbiome-based interventions. Additionally, regulatory bodies must stay informed about emerging research and adapt regulations accordingly to keep pace with rapidly evolving technologies. By establishing clear and comprehensive regulatory frameworks, we can ensure that microbiome-based therapies are developed and utilized in a responsible and effective manner, ultimately benefiting the health and well-being of individuals.

XXIII. GLOBAL VARIATIONS IN HUMAN MICROBIOMES

The global variations in human microbiomes are indicative of the intricate relationship between microbial communities and environmental factors. Research has shown that different populations across the world harbor unique compositions of microbial species in their bodies, influenced by dietary habits, lifestyle choices, genetics, and geographic location. For example, individuals living in urban areas may have a different microbiome profile compared to those in rural settings, reflecting the impact of urbanization on microbial diversity. Moreover, studies have highlighted how cultural practices such as traditional diets and hygiene practices can shape the microbiota composition in distinct populations. Understanding these global variations in human microbiomes is essential for elucidating the complex interplay between microbes and human health on a global scale, emphasizing the need for tailored interventions and personalized medicine approaches that consider the diversity of microbial communities worldwide.

Geographic and Cultural Differences

It is evident that geographic and cultural differences have a significant impact on the composition and diversity of the human microbiome. Studies have shown that individuals from different regions and cultural backgrounds harbor distinct microbial communities within their bodies. Factors such as diet, lifestyle, and environmental exposures vary across different geographic regions and cultures, influencing the types of bacteria that colonize the human gut and other body sites. For example, individuals living in Western societies tend to have a higher abundance of certain Firmicutes species compared to those in non-Western societies, which may contribute to differences in health outcomes such as obesity and metabolic disorders. Understanding these geographic and cultural influences on the human microbiome is paramount in developing personalized approaches to healthcare that take into account individualized microbiome profiles based on geographical and cultural backgrounds. By recognizing these differences, we can better tailor interventions and treatments to optimize health outcomes for diverse populations.

Implications for Global Health

The implications of understanding the human microbiome extend far beyond individual health, with significant impacts on global health initiatives. By comprehending the intricate relationship between the microbiome and human health, researchers and healthcare professionals can develop innovative approaches to address a wide range of health issues on a global scale. For instance, targeting specific bacterial communities within the microbiome could lead to the development of tailored probiotics to combat diseases that are prevalent in certain regions. Moreover, exploring the role of the microbiome in infectious diseases could provide crucial insights into the prevention and treatment of epidemics. With a deeper understanding of how the microbiome influences health outcomes, public health interventions can be designed to promote microbial diversity and support overall well-being on a global level. Ultimately, prioritizing research on the human microbiome has the potential to revolutionize approaches to public health and improve health outcomes worldwide.

Strategies for Cross-Cultural Research

In conducting cross-cultural research within the realm of the human microbiome, it is essential to employ strategies that are sensitive to the diverse cultural backgrounds of the populations involved. One key strategy is ensuring cultural competence among research team members, which involves understanding the beliefs, values, and practices of different cultures to avoid unintentional bias or misinterpretation of findings. Additionally, utilizing a collaborative approach with local researchers and community members can help to foster trust and ensure that the research is culturally relevant and respectful. It is also crucial to employ culturally appropriate data collection methods, such as using language interpreters or adapting survey instruments to align with cultural norms. By implementing these strategies, researchers can enhance the validity and reliability of their cross-cultural studies within the human microbiome field, ultimately contributing to a more comprehensive understanding of the impact of microbial diversity on human health.

XXIV. THE ROLE OF GENETICS IN THE MICROBIOME

Recent research has highlighted the significant role that genetics play in shaping the composition and function of the human microbiome. Genetic variations in both the host and the microbial community can influence the diversity and stability of the microbiome. For example, certain genetic traits in the host can affect the abundance of specific bacterial species in the gut, leading to potential alterations in overall health. On the other hand, variations in the genetic makeup of bacteria themselves can impact their ability to colonize certain niches within the body and interact with host cells. Understanding the interplay between genetics and the microbiome is crucial for unraveling the complexities of human health and disease. By elucidating the genetic factors that contribute to microbiome composition and function, researchers can pave the way for personalized therapies that target the microbiome to promote better health outcomes.

Genetic Influences on Microbiome Composition
Recent research has shed light on the intricate interplay between genetic factors and the composition of the human microbiome. It is now well-established that individual genetic variations can significantly influence the diversity and abundance of microbial species inhabiting our bodies. These genetic influences can shape the microbial communities in various body sites, such as the gut, skin, and oral cavity, impacting their overall balance and functionality. For instance, studies have highlighted specific

genetic loci that are associated with alterations in gut microbiota composition, which can, in turn, affect an individual's susceptibility to certain diseases. Understanding the genetic underpinnings of microbiome composition not only provides insights into the complex mechanisms driving microbial colonization but also opens up new avenues for personalized medicine and targeted interventions to modulate the microbiome for improved health outcomes. The interaction between genetics and the microbiome represents a fascinating frontier in biomedical research, with profound implications for human health and disease.

Personalized Medicine Approaches
Advancements in technology have paved the way for personalized medicine approaches to target specific variations in the human microbiome. By understanding the unique composition of an individual's microbial community, tailored interventions can be designed to optimize health outcomes. This personalized approach takes into account the diversity and complexity of microbial populations within each person, allowing for precision in diagnosis and treatment. Through the use of genomic sequencing, metabolomics, and other cutting-edge tools, researchers can identify the key players in the microbiome that influence various health conditions. This individualized treatment strategy holds great promise in addressing complex diseases and conditions that have proven challenging to manage with traditional approaches. With personalized medicine, the potential for more targeted therapies and improved patient outcomes in the realm of human microbiome research is substantial.

Future Research Directions in Genomics and Microbiomes
Moving forward, future research in genomics and microbiomes holds great potential for advancing our understanding of how bacterial communities within the body influence health outcomes. One direction for future investigation could involve exploring the role of the microbiome in personalized medicine, tailoring treatments based on an individual's unique bacterial composition. This personalized approach may lead to more effective therapies and better outcomes for various diseases. Additionally, further research could delve into the impact of the microbiome on mental health and neurological disorders, shedding light on the intricate connection between gut health and brain function. Understanding these relationships could pave the way for innovative treatments targeting the microbiome to improve mental well-being. Overall, the field of genomics and microbiomes presents a vast landscape for exploration, with the potential to revolutionize healthcare practices and interventions in the coming years.

XXV. TECHNOLOGICAL INNOVATIONS IN MICROBIOME RESEARCH

Technological innovations have significantly advanced the field of microbiome research, providing researchers with powerful tools to study the complex interactions between the human body and its resident microbial communities. High-throughput sequencing techniques, such as metagenomics and metatranscriptomics, have revolutionized our ability to characterize the composition and function of the microbiome in ways that were previously unimaginable. These technologies allow for a comprehensive analysis of the genetic material present in microbial communities, offering insights into the diversity of microorganisms, their gene expression patterns, and their impact on human health. Furthermore, advances in bioinformatics have facilitated the integration of large-scale microbiome data sets, enabling researchers to identify key microbial signatures associated with various diseases and conditions. By harnessing the power of these cutting-edge technologies, scientists are well-positioned to unravel the complex relationship between the human microbiome and health, paving the way for innovative therapeutic interventions and personalized medicine strategies.

New Tools and Techniques
Emerging research in the field of microbiome science is yielding new tools and techniques to explore the complex ecosystems within our bodies. Metagenomic sequencing has revolutionized our understanding of the microbial communities residing in different anatomical sites, allowing for a more comprehensive analysis of their composition and function. The integration of

multi-omics approaches, combining data from genomics, transcriptomics, proteomics, and metabolomics, has provided unprecedented insights into the intricate relationships between host and microbiota. Furthermore, advances in bioinformatics and computational modeling have enabled the development of predictive models to assess the impact of microbial communities on human health. These new tools and techniques offer exciting opportunities for unraveling the role of the human microbiome in various physiological processes and diseases, paving the way for personalized diagnostics and therapeutics tailored to individual microbial signatures. Through these innovative approaches, we are poised to unlock the full potential of the human microbiome in promoting health and preventing disease.

Impact on Research Efficiency and Accuracy

Furthermore, the human microbiome's impact on research efficiency and accuracy cannot be overstated. By studying the vast array of bacteria residing in our bodies, researchers are able to uncover valuable insights into numerous health conditions and diseases, leading to more targeted and effective treatments. The microbiome's influence on immune response, metabolism, and neurological functions has opened up new avenues for research, allowing for a deeper understanding of the intricate interplay between our bodies and the microbial world within us. Moreover, advancements in technology, such as high-throughput sequencing and bioinformatics tools, have revolutionized how microbiome research is conducted, enabling researchers to analyze large datasets quickly and accurately. This increased efficiency in data processing has significantly accelerated the pace of microbiome research, ultimately leading to more precise and reliable findings that can have a profound impact on human health.

Future Technological Trends

One of the most intriguing aspects of the human microbiome is the potential impact of future technological trends on our understanding of these complex microbial communities. As advancements in gene sequencing technologies continue to improve, researchers are now able to explore the microbiome in unprecedented detail, uncovering the intricate relationships between our bodies and the trillions of bacteria that call us home. In the coming years, the integration of artificial intelligence and machine learning algorithms may further enhance our ability to analyze massive amounts of microbiome data, enabling more personalized approaches to healthcare and disease prevention. Additionally, the development of novel probiotics and microbiome-based therapies holds promise for treating a wide range of conditions, from gastrointestinal disorders to mental health issues. By embracing these cutting-edge technologies, we may unlock a deeper understanding of how the human microbiome influences our health and revolutionize the way we approach personalized medicine.

XXVI. FUTURE THERAPEUTIC POTENTIALS OF THE MICROBIOME

The future therapeutic potentials of the microbiome hold promise for revolutionizing the way we approach health and disease. As research in this field continues to advance, we are uncovering the intricate relationships between the microbiome and various health conditions, including autoimmune disorders, gastrointestinal diseases, and even mental health issues. The ability to manipulate the composition of the microbiome through probiotics, prebiotics, and fecal microbiota transplants opens up new avenues for targeted and personalized treatments. By harnessing the therapeutic potential of the microbiome, we may be able to develop novel interventions that could potentially have far-reaching impacts on human health. As we delve deeper into understanding the complex interplay between microbial communities and the human body, the future of medicine may be shaped by our ability to harness the power of the microbiome to prevent, diagnose, and treat a wide range of diseases.

Emerging Therapeutic Techniques
Recent advancements in therapeutic techniques have shown promising results in harnessing the power of the human microbiome to improve health outcomes. One such emerging approach involves fecal microbiota transplantation (FMT), where healthy gut bacteria from a donor are introduced into the gastrointestinal tract of a recipient to restore microbial balance. This method has been particularly successful in treating conditions such as Clostridium difficile infection and inflammatory bowel disease. Additionally, the development of personalized

probiotics tailored to an individual's unique microbiome profile holds great potential in treating a range of health issues. By utilizing cutting-edge technologies like metagenomics and artificial intelligence, researchers can identify specific bacterial strains that are beneficial for a person's health and create targeted therapies. These innovative therapeutic techniques represent a promising frontier in medicine, offering new avenues for improving human health through the manipulation of our microbial inhabitants.

Challenges in Therapeutic Application

One of the significant challenges in therapeutic application of the human microbiome lies in the complexity and diversity of the microbial communities that inhabit different individuals. Each person has a unique microbiome composition, influenced by various factors such as genetics, diet, environment, and lifestyle. This personalized aspect makes it challenging to develop universal treatments that can effectively target specific health issues. Furthermore, the dynamic nature of the microbiome, which can shift in response to various stimuli, adds another layer of complexity. This means that therapeutic approaches need to consider not only the current state of the microbiome but also potential changes over time. Additionally, the interactions between different microbial species and host cells within the body are intricate and still not fully understood, making it difficult to predict how interventions may affect overall health. Overcoming these challenges will require a deeper understanding of the microbiome's role in health and disease, as well as innovative strategies for personalized therapeutic interventions tailored to individual microbiome profiles.

Predictions for Future Therapies

As research on the human microbiome continues to advance, predictions for future therapies are becoming increasingly optimistic. The ability to manipulate the composition of gut bacteria holds great promise for treating a wide range of diseases, from gastrointestinal disorders to neurological conditions. One key focus for future therapies is the development of personalized probiotics tailored to an individual's unique microbiome profile. These probiotics could help restore balance to the gut microbiota and alleviate symptoms of various ailments. Additionally, the use of fecal microbiota transplants (FMT) is an emerging treatment option that has shown remarkable success in treating conditions such as Clostridium difficile infection. As our understanding of the microbiome deepens, we can expect to see a shift towards more targeted and effective therapies that harness the power of our body's own microbial inhabitants to promote health and well-being.

XXVII. MICROBIOME AND PERSONALIZED MEDICINE

The integration of the microbiome into personalized medicine marks a significant advancement in healthcare. By understanding the individual microbial composition of each person, tailored treatments can be developed to target specific health conditions. This personalized approach takes into account the diversity of microbiota in each individual, recognizing that what works for one person may not work for another. This shift towards personalized medicine acknowledges the complexity of the human microbiome and the important role it plays in disease onset and progression. Utilizing microbial data to inform treatment plans allows for more effective and precise interventions, ultimately leading to improved patient outcomes. As we continue to unravel the intricate interactions between our microbiota and health, personalized medicine holds great promise in revolutionizing the way we approach healthcare, offering new possibilities for targeted therapies and interventions tailored to individual microbial profiles.

Tailoring Treatments Based on Microbiome
Understanding the intricate relationship between the human microbiome and our health opens up a realm of possibilities for personalized medicine. Tailoring treatments based on an individual's unique microbiome profile holds the key to more effective and targeted therapies. By analyzing the composition and function of the microbiota, healthcare providers can design interventions that specifically address imbalances or dysbiosis, leading to better outcomes for patients. This approach not only

maximizes the efficacy of treatments but also minimizes potential side effects by working with the body's natural ecosystem. Moreover, as research in this field continues to advance, we are uncovering new insights into the intricate ways in which the microbiome influences disease development and progression. By harnessing this knowledge, we can revolutionize the way we approach healthcare, moving towards a more personalized and holistic model that takes into account the individual's microbiome composition.

Challenges in Implementing Personalized Approaches

The implementation of personalized approaches in healthcare poses several challenges that must be addressed to maximize their effectiveness. One major hurdle is the lack of standardized protocols for personalized medicine, leading to inconsistencies in patient care and outcomes. Additionally, the integration of complex data from various sources, such as genetic, environmental, and lifestyle factors, presents logistical and technological challenges in creating tailored treatment plans. Another obstacle is the need for healthcare providers to undergo specialized training to interpret and apply personalized medicine approaches accurately. Furthermore, issues of data privacy, security, and ethical considerations must be carefully navigated to ensure patient confidentiality and autonomy. Despite these challenges, the potential benefits of personalized medicine in improving patient outcomes and reducing healthcare costs are substantial, warranting continued efforts to overcome implementation barriers and advance this innovative approach in healthcare practice.

Future of Personalized Medicine and Microbiome

With the rapid advancements in technology and research, the future of personalized medicine in relation to the human microbiome holds great promise. By leveraging the information obtained from analyzing individual microbiomes, healthcare providers can tailor treatment plans to specific patients, taking into account their unique microbial composition. This personalized approach can potentially lead to more effective and targeted therapies, minimizing the risk of adverse reactions and improving overall patient outcomes. Furthermore, as our understanding of the microbiome continues to deepen, we may uncover new ways in which these microorganisms influence various aspects of human health, opening up avenues for novel therapeutic interventions. As we move forward, integrating personalized medicine with insights from the microbiome has the potential to revolutionize healthcare practices, providing patients with more precise and individualized treatments based on their microbial profile.

XXVIII. MICROBIOME IN NON-HUMAN MODELS

Recent studies have highlighted the importance of investigating the microbiome in non-human models to better understand its impact on human health. By utilizing animal models such as mice, zebrafish, and fruit flies, researchers can study the intricate interactions between host genetics, diet, and the microbiome. These models provide valuable insights into how specific microbial communities influence various physiological processes, such as immune development, metabolism, and neurological function. Furthermore, non-human models allow for controlled experimental conditions that are not feasible in human studies, enabling researchers to manipulate the microbiome and observe its effects on host physiology. These findings can then be extrapolated to humans, shedding light on potential therapeutic targets for a range of health conditions. Therefore, exploring the microbiome in non-human models is essential for advancing our understanding of the complex relationship between microbial communities and host health.

Studies on Animal Microbiomes
Recent studies on animal microbiomes have provided valuable insights into the intricate relationship between the host and their microbial inhabitants. This emerging field has shed light on the diverse composition of microbiota in various animal species, influencing both their physiology and ecological interactions. Through sophisticated analytical techniques, researchers have unraveled the complex network of microbial communities resid-

ing in the gut, skin, and other body parts of animals. These studies have not only deepened our understanding of host-microbe interactions but also highlighted the significant impact of microbiomes on animal health, immunity, and behavior. By exploring the microbial diversity within different animal species, scientists are uncovering novel therapeutic strategies to improve animal health and well-being. As we delve further into the realm of animal microbiomes, we are poised to unlock even more secrets of this intricate symbiotic relationship between animals and their microbial counterparts.Insights Gained and Their Relevance to Humans Recent research on the human microbiome has provided invaluable insights into the complex relationship between our bodies and the trillions of bacteria that reside within us. The role of the gut microbiota in modulating immune responses, nutrient absorption, and even neurological functions has opened up new avenues for understanding human health and disease. By uncovering the intricate mechanisms by which these microbes interact with our bodies, scientists have gained a deeper appreciation for the essential role that they play in maintaining homeostasis and promoting overall well-being. Furthermore, these insights have practical relevance for human health, as they have the potential to guide the development of novel therapeutic strategies for treating a wide range of conditions, from gastrointestinal disorders to mental health issues. Understanding the ways in which our microbiota influences our health can lead to personalized interventions that target the microbiome, offering new opportunities for promoting better health outcomes in diverse populations.

Ethical Considerations in Animal Studies

Furthermore, ethical considerations in animal studies are of paramount importance when conducting research on the human microbiome. While these studies provide valuable insights into the potential benefits of probiotics and microbial therapies, it is crucial to ensure that the well-being of animals involved is not compromised. Researchers must adhere to strict ethical guidelines to minimize harm and uphold the principles of animal welfare. This involves obtaining informed consent, minimizing discomfort and distress, and utilizing the fewest animals possible to achieve research objectives. Additionally, researchers must consider alternative methods such as in vitro studies or computational models to reduce the reliance on animal experimentation. By prioritizing ethical considerations in animal studies, researchers can conduct meaningful research on the human microbiome while upholding moral values and respecting the rights of all living beings involved.

XXIX. COMMERCIALIZATION OF MICROBIOME RESEARCH

As microbiome research continues to advance, the commercialization of this field presents both opportunities and challenges. With the growing understanding of how the microbiome influences human health, there is a surge in interest from pharmaceutical companies, biotech firms, and even beauty companies looking to capitalize on the potential benefits of manipulating the microbiome. This influx of commercial interest brings new funding opportunities for research, which can lead to groundbreaking discoveries and advancements in medical treatments. However, commercialization also raises concerns about conflicts of interest, ethical implications, and the potential for profit-driven decisions to overshadow scientific integrity. As the commercial landscape of microbiome research evolves, it is crucial for researchers, policymakers, and industry stakeholders to navigate these complexities carefully to ensure that the potential benefits of microbiome-based products and therapies are realized without sacrificing the integrity and credibility of scientific research.

Current Market Trends
The dynamic landscape of current market trends in the field of human microbiome research reflects a growing recognition of the importance of microbial communities in shaping human health. One key trend is the increasing investment in microbiome-based therapies by pharmaceutical companies, highlighting a shift towards personalized medicine that takes into ac-

count individual variations in microbial composition. Additionally, the burgeoning consumer interest in probiotics and prebiotics signifies a growing awareness of the role of gut bacteria in maintaining overall well-being. Moreover, the rise of direct-to-consumer microbiome testing services points towards a democratization of access to information about one's own microbial profile. These trends underscore a paradigm shift towards a more holistic understanding of human health, emphasizing the intricate interplay between the microbiome and the host. As research in this field progresses, it is essential to critically evaluate these market trends to ensure that they align with the goal of promoting health and wellness through microbial interventions.

Ethical and Practical Challenges

The exploration of the human microbiome poses various ethical and practical challenges that must be carefully considered in research and medical applications. Ethically, the use of human subjects in microbiome studies raises concerns about informed consent, privacy, and potential harm to participants. With the vast diversity of microbial communities in different individuals, issues of data ownership and protection also come to the forefront. Moreover, the practical challenges include the complexity of analyzing and interpreting microbiome data, as well as the need for standardized protocols and methodologies. The dynamic nature of the microbiome and its interactions with environmental factors further complicate research efforts. Despite these challenges, addressing ethical considerations and overcoming practical obstacles are essential for advancing our understanding of how the human microbiome influences health outcomes and developing targeted interventions to improve human health.

Future Market Predictions

As we look towards the future of the market, it is evident that advancements in technology and research will continue to drive growth and innovation in the field of human microbiome research. With ongoing collaborations between scientists, healthcare providers, and biotech companies, we can expect to see a wave of new products and therapies targeting microbiome-related conditions. The increasing interest in personalized medicine and precision healthcare will likely lead to a surge in demand for microbiome testing services and tailored treatments. Moreover, the rising awareness of the microbiome's impact on various aspects of health, from immunity to mental well-being, suggests that consumer interest in microbiome-friendly products will also rise. This anticipated growth in the microbiome market signifies a shift towards a more holistic approach to healthcare, where understanding the intricacies of the human microbiome plays a central role in shaping the future of medicine and wellness.

XXX. PARTNERSHIPS AND COLLABORATIONS IN MICROBIOME RESEARCH

One significant aspect of advancing our understanding of the human microbiome lies in the formation of partnerships and collaborations within the field of microbiome research. By coming together, researchers and scientists can combine their expertise, resources, and data to tackle complex questions and challenges related to the microbiome. These partnerships enable the sharing of knowledge and technologies, leading to more robust and comprehensive studies that can provide deeper insights into the intricate interactions between the microbiome and human health. Moreover, collaborations between academia, industry, and healthcare institutions can facilitate the translation of research findings into practical applications, such as the development of new therapies or diagnostic tools. Through these synergistic efforts, the field of microbiome research can continue to expand and contribute to improving human health outcomes.

Role of Interdisciplinary Collaboration

Interdisciplinary collaboration is essential in the study of the human microbiome due to its complex and multifaceted nature. By bringing together experts from various fields such as microbiology, immunology, genetics, and medicine, researchers can gain a more comprehensive understanding of how the microbiome influences our health. For example, microbiologists can analyze the composition of bacterial communities in the body, while immunologists can investigate how these microbes interact with the immune system. Geneticists can identify specific genes that are associated with certain microbial populations, and clinicians can apply this knowledge to develop personalized treatments for patients. Through interdisciplinary collaboration, researchers can bridge gaps in knowledge, uncover new insights, and ultimately advance our understanding of the intricate relationship between the human microbiome and health. This collaborative approach is crucial in unraveling the complexities of the microbiome and its impact on human health.

Major Collaborative Projects

The success of major collaborative projects in the field of microbiome research hinges on the effective coordination of multidisciplinary teams with diverse expertise. These projects often require a harmonious integration of knowledge from microbiologists, immunologists, geneticists, bioinformaticians, and clinicians to unravel the complex interactions within the human microbiome. By pooling resources, skills, and data, these collaborative efforts can generate comprehensive datasets that offer a more holistic understanding of the microbiome's impact on human health. Moreover, such collaboration fosters innovation by encouraging cross-pollination of ideas and methodologies, leading to groundbreaking discoveries that would be unattainable through individual research endeavors. Embracing the collaborative spirit in microbiome research not only accelerates scientific progress but also promotes a more nuanced and inclusive approach to addressing the intricate relationship between the microbiome and human health.

Benefits and Challenges of Collaboration

Collaboration among researchers in the field of human microbiome studies presents numerous benefits as well as challenges.

One of the key advantages of collaboration is the ability to pool together a diverse range of expertise and resources, leading to more comprehensive and innovative research outcomes. By combining different perspectives and skill sets, researchers can tackle complex scientific questions more effectively and potentially make groundbreaking discoveries. Additionally, collaboration can help foster a sense of camaraderie and shared purpose among scientists, creating a stimulating and supportive environment for conducting research. However, collaboration can also present challenges, such as coordinating efforts across multiple institutions or overcoming communication barriers between team members. Ensuring effective teamwork, managing conflicts of interest, and maintaining data integrity are all critical considerations when engaging in collaborative research efforts in the field of human microbiome studies. Ultimately, successful collaboration can significantly advance our understanding of the role of bacteria in the body and its impact on human health.

XXXI. FUNDING AND INVESTMENT IN MICROBIOME RESEARCH

In the realm of microbiome research, securing adequate funding and investment is essential to drive progress and innovation in this burgeoning field. The complexity and diversity of the human microbiome present a myriad of opportunities for groundbreaking discoveries, yet the high costs associated with conducting rigorous research continue to pose a significant challenge. In order to unlock the full potential of microbiome science, increased financial support from government agencies, private organizations, and philanthropic foundations is imperative. By investing in microbiome research, stakeholders can pave the way for novel therapeutics, diagnostics, and interventions that have the potential to revolutionize healthcare practices and improve patient outcomes. Moreover, fostering collaboration between researchers, industry partners, and funding bodies can further accelerate the translation of microbiome findings into tangible benefits for human health. Ultimately, sustained investment in microbiome research holds tremendous promise for unraveling the intricate relationships between our microbial inhabitants and overall well-being.

Overview of Funding Sources

One critical aspect to consider when exploring the human microbiome is the various sources of funding that support research in this field. Funding for microbiome research comes from a variety of sources, including government agencies, non-profit organizations, private foundations, and industry partners. Government agencies such as the National Institutes of Health (NIH) and the National Science Foundation (NSF) are major contributors to funding in this area, providing grants to support basic and translational research. Non-profit organizations like the Bill & Melinda Gates Foundation and the Wellcome Trust also play a significant role in funding microbiome research, focusing on global health and infectious diseases. Furthermore, industry partnerships with pharmaceutical companies and biotechnology firms can provide additional financial support for cutting-edge research and development. Overall, the diverse range of funding sources ensures that the study of the human microbiome remains a dynamic and growing field with the potential to revolutionize healthcare practices and improve patient outcomes.

Trends in Investment

Recent trends in investment highlight a shift towards sustainable and socially responsible practices. Investors are increasingly considering environmental, social, and governance (ESG) factors when making investment decisions, recognizing the importance of ethical and sustainable practices in the long-term success of companies. This trend is driven by a growing awareness of the impact of businesses on the environment and society, as well as the potential financial risks associated with unsustainable practices. As a result, companies that prioritize ESG considerations are seen as more attractive investments, leading to a rise in sustainable investing strategies. This shift not only reflects changing consumer preferences and regulatory pressures but also indicates a broader recognition of the interplay between financial performance and responsible business practices. As the investment landscape continues to evolve, integrating ESG factors into decision-making processes will likely become the new norm, shaping the future of investment practices.

Impact of Funding on Research Progress

In the realm of scientific inquiry, funding is an essential factor that significantly influences the progress of research endeavors. Adequate funding not only sustains the operations of research projects but also enables researchers to explore innovative ideas and methodologies, leading to scientific breakthroughs. Particularly in the field of human microbiome research, where the complexities of microbial communities within the body pose unique challenges, sufficient funding plays a critical role in advancing our understanding of how these bacteria impact our health. With ample financial support, researchers can conduct extensive studies, invest in cutting-edge technology, and attract top talent to their teams, ultimately accelerating the pace of discovery and opening up new avenues of investigation. Conversely, insufficient funding can hinder progress, stifle innovation, and limit the scope of research projects, preventing researchers from fully realizing the potential benefits of their work. Thus, the impact of funding on research progress in the study of the human microbiome cannot be overstated, underscoring the importance of robust financial support in advancing our knowledge of this crucial aspect of human health.

XXXII. MICROBIOME RESEARCH AND PUBLIC HEALTH POLICY

Recent advancements in microbiome research have shed light on the intricate relationship between the microbes inhabiting our bodies and their impact on our health. As scientists continue to uncover the complexities of the human microbiome, it becomes increasingly evident that these tiny organisms play a pivotal role in shaping our overall well-being. This growing body of knowledge has significant implications for public health policy, as understanding the microbiome can lead to the development of targeted interventions to prevent and treat a myriad of diseases. By incorporating microbiome research into public health initiatives, policymakers can implement strategies that harness the power of beneficial bacteria to promote health and prevent illness. Therefore, bridging the gap between microbiome research and public health policy holds great promise in revolutionizing healthcare practices and improving population health outcomes.

Influence on Health Policy Making

The influence of the human microbiome on health policy making is a topic that has gained significant attention in recent years. As researchers uncover more about the intricate relationship between the bacteria in our bodies and our overall health, policymakers are starting to recognize the importance of integrating this knowledge into healthcare strategies. The microbiome has been linked to a wide range of health conditions, from obesity to autoimmune diseases, highlighting the need for policies that support research into microbiome-based treatments and therapies. By understanding how the microbiome influences our health, policymakers can develop more effective preventive measures and treatments, ultimately leading to improved public health outcomes. Incorporating microbiome research into health policy making has the potential to revolutionize healthcare systems and pave the way for personalized medicine approaches that take into account each individual's unique microbiome composition.

Policy Challenges

Policy challenges in the field of human microbiome research present complex issues that require careful consideration. One of the primary challenges lies in regulating the testing and implementation of microbiome-based therapies. Current regulatory frameworks may not adequately address the unique nature of these treatments, leading to ambiguity in guidelines and potential safety concerns for patients. Another key policy challenge is the ethical implications of manipulating the human microbiome. Questions surrounding informed consent, data privacy, and the potential for unintended consequences must be carefully navigated to ensure the responsible advancement of microbiome-based interventions. Additionally, disparities in access to microbiome research and therapies highlight the need for policies that promote equitable distribution of resources and benefits. Addressing these policy challenges will be essential in fostering beneficial advancements in human microbiome research while safeguarding patient safety and maintaining ethical standards.

Recommendations for Policymakers

In light of the growing recognition of the importance of the human microbiome in influencing health, policymakers must prioritize funding and support for research initiatives aimed at further exploring this intricate relationship. Collaborative efforts between scientists, healthcare professionals, and policymakers are essential to translating research findings into tangible public health interventions that promote microbial diversity and resilience. Policies should be devised to encourage the development and implementation of microbiome-targeted therapies, such as probiotics and prebiotics, to optimize health outcomes. Additionally, policymakers should consider incorporating microbiome testing and analysis into routine healthcare practices to personalize treatment plans and improve patient outcomes. By investing in research, fostering interdisciplinary collaborations, and integrating microbiome-focused interventions into healthcare systems, policymakers can address the burgeoning field of microbiome science and harness its full potential to enhance human health and well-being.

XXXIII. MICROBIOME AND ENVIRONMENTAL HEALTH

The intricate relationship between the human microbiome and environmental health is a topic of increasing interest and importance in the field of microbial ecology. The microbiome, comprised of trillions of microorganisms residing in and on our bodies, has been shown to have significant implications for our overall well-being. From influencing the immune system to impacting metabolism, these bacterial communities play a crucial role in maintaining homeostasis within the body. However, recent research has also highlighted the impact of environmental factors on the composition and function of the microbiome. Factors such as diet, pollution, and antibiotic use can disrupt the delicate balance of microbial populations, leading to dysbiosis and potential health issues. Understanding the complex interplay between the microbiome and environmental exposures is essential for developing strategies to protect and promote human health in an increasingly urbanized and industrialized world.

Interaction Between Environmental Factors and Microbiome

One of the key aspects to consider when exploring the human microbiome is the intricate interaction between environmental factors and the microbiome. The microbiome, composed of trillions of microbes residing in and on the human body, is highly sensitive to its surroundings. Environmental factors such as diet, lifestyle, exposure to pollutants, and even geographical location can significantly impact the composition and function of the microbiome. For instance, a diet rich in fiber can promote the growth of beneficial bacteria in the gut, while exposure to antibiotics can disrupt the delicate balance of microbial communities. Understanding how these environmental factors shape the microbiome is crucial in elucidating the mechanisms by which the microbiome influences human health. By studying the dynamic interplay between environmental factors and the microbiome, we can gain valuable insights into how to modulate the microbiome to optimize health outcomes and prevent various diseases.

Impact on Public Health

The impact of the human microbiome on public health is a topic of growing interest and significance in the field of medicine. Emerging research has shown that the microbiota residing in our bodies play a crucial role in maintaining our health and influencing various disease states. The interactions between the host and these microbial communities have significant implications for conditions such as obesity, diabetes, and even mental health disorders. Understanding the intricate balance of the microbiome can lead to novel therapeutic interventions that target specific bacterial species to improve health outcomes. Moreover, research on the microbiome has the potential to revolutionize personalized medicine, allowing for tailored treatments based on an individual's unique microbial profile. By elucidating the role of the microbiome in public health, we can pave the way for innovative strategies to prevent and treat a wide range of diseases, ultimately improving the overall well-being of populations worldwide.

Strategies for Environmental Management

In the realm of environmental management, strategies play a vital role in preserving and protecting our ecosystems. One effective approach is the implementation of sustainable practices that minimize negative impacts on the environment while promoting long-term ecological balance. This can involve the adoption of renewable energy sources, waste reduction, and recycling programs to minimize the carbon footprint. Additionally, integrating biodiversity conservation efforts into land management practices can help maintain healthy ecosystems and prevent species extinction. Collaborative initiatives involving government agencies, businesses, and communities can also lead to more comprehensive and effective environmental management strategies. By fostering a culture of environmental stewardship and promoting awareness of the importance of conservation, we can work towards a more sustainable future for generations to come. Through proactive and strategic planning, we can mitigate environmental degradation and ensure a healthier planet for all living organisms.

XXXIV. CHALLENGES IN MICROBIOME SAMPLE COLLECTION AND STORAGE

As we delve deeper into exploring the intricacies of the human microbiome, it becomes evident that challenges in microbiome sample collection and storage pose significant hurdles in research endeavors. The collection of microbiome samples involves ensuring the preservation of bacterial diversity and viability while minimizing external contamination. Issues such as sample handling, transport conditions, and storage methods can greatly impact the quality and reliability of the data obtained. Moreover, the variability in sample collection protocols across studies can lead to inconsistencies and difficulties in comparing results. Standardization of collection procedures and the development of robust storage techniques are imperative to enhance the reproducibility and reliability of microbiome research. By addressing these challenges, researchers can gain a more comprehensive understanding of the human microbiome and its impact on health, paving the way for advancements in personalized medicine and targeted microbial therapies.

Best Practices for Sample Collection
Research has shown that the collection of samples is crucial in studying the human microbiome. Best practices for sample collection include minimizing contamination by using sterile collection tools and ensuring proper storage and transportation conditions to maintain the integrity of the samples. It is also important to standardize collection methods across studies to facilitate comparisons and reproducibility. Additionally, obtaining samples from multiple sites within the body can provide a more

comprehensive understanding of the microbiome composition and its impact on health. By following these best practices, researchers can gather high-quality data that will contribute to a more accurate depiction of the complex microbial communities that inhabit the human body. This attention to detail in sample collection is essential for advancing our knowledge of the human microbiome and its role in health and disease.

Storage and Preservation Techniques

Research on storage and preservation techniques for human microbiome samples is crucial in maintaining the integrity and quality of the collected data. Various methods have been developed to ensure the stability of the microbial content within samples, such as freezing at ultra-low temperatures or using preservatives to prevent microbial growth. These techniques are essential for long-term storage and future analysis of the microbiome composition. Careful consideration must be taken in choosing the most suitable method based on the specific research goals and sample characteristics. The choice of storage and preservation technique can greatly impact the accuracy and reliability of the data obtained, making it a critical component in microbiome research. By employing proper storage and preservation techniques, researchers can ensure that the microbiome samples remain representative of the in vivo microbial community, allowing for meaningful analysis and interpretation of the data.

Impact on Research Quality

The impact of the human microbiome on research quality is undeniable. By understanding the complex interactions between the trillions of bacteria residing in our bodies and their influence on our health, researchers are able to uncover novel insights that can lead to the development of groundbreaking therapies and interventions. Through advanced technologies such as metagenomics and high-throughput sequencing, scientists are now able to explore the vast diversity of microbial species within the human microbiome with unprecedented depth and precision. This has opened up new avenues for studying the role of the microbiome in various diseases, such as autoimmune disorders, obesity, and even mental health conditions. By unraveling the intricate mechanisms by which these bacteria impact our physiological processes, researchers can enhance the quality and depth of their studies, ultimately leading to more targeted and effective healthcare strategies. In conclusion, the study of the human microbiome not only expands our understanding of human biology but also elevates the quality of research in the field of medicine.

XXXV. DATA ANALYSIS AND INTERPRETATION IN MICROBIOME RESEARCH

The field of microbiome research has seen significant advancements in recent years, particularly in the realm of data analysis and interpretation. As researchers delve deeper into the complex microbial communities that inhabit our bodies, innovative techniques and technologies have been developed to handle the vast amount of data generated by microbiome studies. Through sophisticated bioinformatics tools and statistical analyses, scientists are able to identify patterns, relationships, and correlations within the microbiome data, shedding light on the intricate interactions between microbes and their human hosts. This in-depth data analysis not only provides valuable insights into the composition and function of the microbiome but also allows researchers to draw meaningful conclusions about how these microbial communities influence human health and disease. By leveraging sophisticated data analysis techniques, microbiome researchers are uncovering groundbreaking discoveries that have the potential to revolutionize our understanding of the role bacteria play in shaping our overall well-being.

Advanced Analytical Techniques

The integration of advanced analytical techniques in studying the human microbiome has revolutionized our understanding of the intricate interactions between bacteria and human health. Techniques such as metagenomics, metatranscriptomics, and metabolomics allow researchers to explore the complex microbial communities that reside in our bodies with unprecedented depth and precision. By analyzing the genetic material, gene expression, and metabolites produced by these microbes, scientists can unravel the intricate mechanisms through which they influence our health. For instance, metagenomic analysis can identify specific microbial species associated with certain diseases, while metabolomic profiling can reveal the metabolic pathways involved in host-microbe interactions. These techniques not only provide valuable insights into the microbial composition of the human body but also shed light on the underlying mechanisms driving the symbiotic relationship between bacteria and their human hosts. Ultimately, the application of advanced analytical techniques in microbiome research holds the key to unlocking new therapeutic strategies for promoting health and combating disease.

Challenges in Data Interpretation
Interpreting data related to the human microbiome presents challenges that must be addressed to fully understand its impact on health. One such challenge is the complexity and diversity of microbial communities within the human body. With trillions of bacteria residing in various niches, deciphering the interactions between different species and their effects on the host can be daunting. Additionally, the dynamic nature of the microbiome makes it difficult to establish causality in studies, as correlations may not always indicate direct relationships. Furthermore, standardizing methodologies for data collection and analysis is crucial to ensure consistency and reproducibility across studies. Without clear guidelines and protocols, comparisons between different research findings become challenging. Overcoming these challenges in data interpretation is essential

to advancing our knowledge of how the human microbiome influences health and disease, paving the way for targeted interventions and personalized medicine approaches.

Improving Accuracy and Reliability

Research in the field of human microbiome analysis has made significant strides in recent years, but one key area that requires further attention is the improvement of accuracy and reliability in the data collected. In order to draw meaningful conclusions about the role of bacteria in the body and their impact on health, it is crucial that data is as precise and consistent as possible. This can be achieved through the use of standardized methods for sample collection, processing, and analysis, as well as the implementation of rigorous quality control measures. By ensuring that data is accurate and reliable, researchers can have greater confidence in their findings and the conclusions drawn from them. Ultimately, this will lead to a deeper understanding of the complex interactions between the human microbiome and our health, paving the way for more targeted and effective interventions in the future.

XXXVI. MICROBIOME AND INFECTIOUS DISEASES

The intricate relationship between the human microbiome and infectious diseases is a multifaceted and dynamic field of study. The microbiome, composed of trillions of microorganisms residing in our body, plays a pivotal role in modulating our immune system and defending against pathogens. However, disruptions in the balance of these microbial communities can lead to susceptibility to infections. Pathogenic bacteria can outcompete beneficial microbes, causing dysbiosis and creating an environment conducive to infections. Furthermore, certain pathogenic bacteria have evolved mechanisms to manipulate host immune responses and evade clearance, contributing to the persistence of infectious diseases. Understanding the interplay between the microbiome and infectious diseases is paramount in developing innovative therapeutic strategies, such as probiotics and fecal microbiota transplantation, to restore microbial homeostasis and combat infections effectively. By unraveling the complexities of the microbiome, we can potentially revolutionize the management and treatment of various infectious diseases.

Role in Disease Prevention
The role of the human microbiome in disease prevention is a complex and multifaceted one. Research has shown that the bacteria residing in our bodies can influence our immune system, metabolism, and overall health. By maintaining a diverse and balanced microbiome, we can help prevent various diseases, including autoimmune disorders, obesity, and gastrointestinal conditions. The gut microbiota, in particular, plays a crucial role

in maintaining our overall health by promoting digestion, nutrient absorption, and maintaining the gut barrier function. Dysbiosis, or an imbalance in the microbiome, has been associated with various diseases, highlighting the importance of maintaining a healthy microbiome for disease prevention. Understanding the intricate relationship between our microbiome and disease prevention can pave the way for new therapeutic strategies that harness the potential of our bacterial inhabitants to promote health and prevent illness.

Microbiome's Influence on Pathogen Dynamics

The intricate relationship between the human microbiome and pathogen dynamics is a complex and dynamic interplay that can have significant implications for human health. Recent research has demonstrated that the composition and diversity of the microbiome can impact the colonization and virulence of pathogens within the host. For example, certain commensal bacteria can compete with pathogenic species for resources and space, thereby reducing the likelihood of infection. Additionally, the microbiome can modulate the host immune response, influencing the ability of pathogens to establish themselves and cause disease. Understanding these interactions is crucial for developing new strategies to prevent and treat infections. By elucidating the mechanisms by which the microbiome influences pathogen dynamics, we can potentially harness the power of these microbial communities to promote health and mitigate the impact of infectious diseases.

Strategies for Infectious Disease Management

In the realm of infectious disease management, various strategies have been developed to combat the spread of illnesses caused by pathogens. One key approach is through vaccination, which helps build immunity against specific diseases and reduce the likelihood of infection within a population. Additionally, public health measures such as quarantine, isolation, and contact tracing are essential in controlling the spread of infectious diseases. Education and communication campaigns are also crucial in increasing awareness about the importance of hand hygiene, respiratory etiquette, and vaccination. Furthermore, antimicrobial stewardship programs help prevent the development of antibiotic resistance by promoting the judicious use of antibiotics. By employing a comprehensive and multi-faceted approach to infectious disease management, we can effectively reduce the burden of infectious diseases and safeguard public health.

XXXVII. MICROBIOME AND RESISTANCE TO ANTIBIOTICS

The intricate relationship between the human microbiome and resistance to antibiotics is a topic of growing concern in the field of medicine. Recent studies have highlighted the role that the microbiome plays in modulating the effectiveness of antibiotic treatments. It has been observed that certain bacterial species within the microbiome can confer resistance to antibiotics, making it harder to eradicate infections. This phenomenon underscores the importance of understanding the interplay between the microbiome and antibiotic resistance in order to develop more targeted and effective treatment strategies. By investigating the mechanisms by which the microbiome influences antibiotic resistance, researchers can potentially identify new therapeutic targets and approaches to combat this pressing issue. Ultimately, a deeper understanding of the microbiome's impact on antibiotic resistance can lead to improved clinical outcomes and better management of infectious diseases in the future.

Development of Resistance

One of the critical aspects to consider in the context of the human microbiome is the development of resistance. As bacteria interact with various environments within the body, they may evolve mechanisms to resist the effects of antibiotics or other interventions aimed at controlling their growth. This process, known as antibiotic resistance, poses a significant challenge to modern healthcare practices as it limits the effectiveness of treatment options. Understanding the mechanisms and pathways through which bacteria develop resistance is essential for devising strategies to combat this phenomenon. By studying the evolutionary processes that drive resistance development, researchers can potentially identify new targets for intervention and develop novel approaches to managing bacterial infections. Furthermore, investigating how bacteria within the microbiome develop resistance not only sheds light on microbial behavior but also emphasizes the importance of responsible antibiotic use to mitigate the spread of resistant strains and preserve the efficacy of existing treatments.

Strategies to Combat Resistance
The search for effective strategies to combat resistance in the human microbiome presents a multifaceted challenge that requires a comprehensive approach. One key strategy involves promoting the diversity of beneficial bacteria through dietary interventions that support the growth of probiotic strains. Probiotics, such as Lactobacillus and Bifidobacterium, have been shown to enhance gut health and bolster the immune system, thereby reducing the risk of colonization by harmful pathogens. In addition to dietary measures, the judicious use of antibiotics is essential in preventing the proliferation of antibiotic-resistant bacteria in the microbiome. By promoting the development of alternative therapies that target specific pathogens while preserving the balance of the microbiome, researchers can minimize the emergence of resistance and safeguard the delicate ecosystem of microbial communities within the body. This integrated approach holds promise for addressing resistance in the human microbiome and promoting optimal health outcomes.

Future Directions in Research and Treatment
As researchers continue to delve deeper into the intricate world

of the human microbiome, future directions in research and treatment are poised to revolutionize the field of medicine. One area of focus lies in personalized microbiome-based therapies, which aim to tailor treatment strategies to individual patients based on their unique microbial profiles. This targeted approach holds great promise in improving treatment outcomes for various diseases, ranging from autoimmune disorders to mental health conditions. Moreover, advancements in microbiome sequencing technologies are expected to drive the development of novel diagnostics and therapeutics, enabling healthcare providers to better understand and harness the power of the microbiome in promoting health and preventing disease. By integrating cutting-edge research findings with clinical practice, the future of microbiome-based medicine holds immense potential in transforming healthcare delivery and improving patient outcomes.

XXXVIII. LEGAL ASPECTS OF MICROBIOME RESEARCH

Given the increasing interest in the human microbiome and its impact on health, it is essential to consider the legal aspects of microbiome research. As researchers delve deeper into understanding the complex interactions between the human body and its resident microbes, ethical and legal considerations become paramount. Issues such as privacy, informed consent, and data ownership arise when collecting and analyzing microbiome data from human participants. Furthermore, intellectual property rights related to discoveries in microbiome research pose challenges in translating scientific findings into commercial applications. In order to navigate these legal complexities effectively, researchers and institutions must adhere to established guidelines and regulations to ensure the ethical conduct of microbiome research. By developing a robust framework that addresses legal considerations, we can safeguard the rights of individuals participating in microbiome studies while promoting groundbreaking discoveries in this burgeoning field.

Intellectual Property Issues

The issues surrounding intellectual property in the context of the human microbiome are complex and multifaceted. As researchers delve deeper into understanding the intricate relationship between the bacteria in our bodies and our health, questions arise regarding who owns the data generated from microbiome research. With the potential for groundbreaking discoveries and innovative treatments to emerge from this field, the race to secure patents and protect intellectual property rights intensifies.

However, the challenge lies in balancing the need for commercial incentives with the ethical considerations of access to essential health information. By navigating these intellectual property issues thoughtfully and transparently, stakeholders can ensure that the benefits of microbiome research are equitably distributed, fostering collaboration and advancement in this rapidly evolving field of science. In doing so, we can harness the power of the human microbiome to revolutionize healthcare and improve lives globally.

Compliance with International Laws

When considering the human microbiome and its impact on health, compliance with international laws becomes essential in ensuring the ethical treatment of research subjects and the protection of human rights. International laws and guidelines, such as the Declaration of Helsinki, set forth principles for conducting research on human subjects, including obtaining informed consent, minimizing potential harms, and upholding the integrity and confidentiality of data. Compliance with these standards is crucial to maintaining the trust of participants and the scientific community as a whole. Additionally, adherence to international laws helps prevent exploitation and ensures that research is conducted in a fair and transparent manner. By following these guidelines, researchers can uphold the highest ethical standards in studying the human microbiome, thereby contributing to advances in healthcare while respecting the rights and dignity of individuals involved in research studies.

Future Legal Challenges

The future legal challenges surrounding the human microbiome are multifaceted and complex. As research continues to uncover the intricate relationship between our microbiota and overall health, legal frameworks will need to adapt to ensure the protection of individual rights and privacy. One key issue that may arise is the regulation of microbiome-based therapies and treatments, as well as the potential for discrimination based on microbiome profiles. Additionally, questions surrounding ownership and control of microbiome data, particularly in the context of personalized medicine and genetic testing, will need to be addressed. It is imperative that legal systems evolve alongside scientific advancements to safeguard against potential misuse or unethical practices involving the human microbiome. As such, policymakers must collaborate with the scientific community to develop ethical guidelines and regulations that prioritize the well-being and autonomy of individuals in the ever-evolving landscape of microbiome research.

XXXIX. MICROBIOME AND LIFESTYLE FACTORS

A key aspect of the human microbiome that warrants discussion is the significant influence of lifestyle factors on its composition and function. Research has shown that diet, physical activity, stress levels, and medication use can all impact the diversity and balance of microorganisms in our bodies. For instance, a diet rich in fiber promotes the growth of beneficial bacteria in the gut, while excessive consumption of processed foods may lead to dysbiosis and inflammation. Similarly, regular exercise has been linked to a more diverse and resilient microbiome, which in turn can support overall health. Furthermore, stress and certain medications, such as antibiotics, can disrupt the delicate microbial ecosystem within us. Understanding the dynamic interplay between our lifestyle choices and the microbiome is crucial for developing personalized strategies to optimize our health and prevent disease. By considering these factors,
we can harness the power of our microbial companions to enhance well-being and longevity.

Impact of Exercise on Microbiome
Numerous studies have highlighted the significant impact of exercise on the human microbiome. Exercise has been shown to promote a more diverse and abundant microbial community in the gut, which is associated with better overall health. Physical activity can increase the production of short-chain fatty acids, which are essential for maintaining gut health and reducing inflammation. Furthermore, exercise has been linked to the mod-

ulation of certain bacteria that can help protect against conditions such as obesity and diabetes. Regular physical activity also appears to have a positive influence on the immune system by promoting a healthier balance of beneficial bacteria. These findings underscore the intricate relationship between exercise and the microbiome, suggesting that incorporating regular physical activity into one's routine can have profound implications for improving overall health and well-being.

Effects of Stress and Sleep

Recent research has shown a clear connection between stress, sleep, and the human microbiome. The effects of stress on the body can lead to dysbiosis, an imbalance in the gut microbiota, which can result in a weakened immune system and increased susceptibility to various health conditions. Chronic stress has been linked to a decrease in beneficial bacteria and an overgrowth of harmful microbes, contributing to inflammation and oxidative stress in the body. Additionally, inadequate sleep can also disrupt the balance of the microbiome, as it plays a crucial role in regulating the circadian rhythm and immune function. Poor sleep quality has been associated with a decrease in microbial diversity and increased levels of potentially harmful bacteria in the gut. Overall, the interplay between stress, sleep, and the microbiome highlights the importance of maintaining a healthy lifestyle to support the balance of beneficial bacteria in the body and promote overall well-being.

Lifestyle Modifications for Optimal Microbiome Health

Emerging research suggests that lifestyle modifications can significantly impact the composition and diversity of the microbiome, ultimately influencing our overall health. One key lifestyle factor is diet; a diverse range of prebiotic-rich foods such as fruits, vegetables, whole grains, and legumes can promote the growth of beneficial bacteria in the gut. Additionally, avoiding processed foods high in sugar and unhealthy fats can help maintain a healthy microbiome. Regular physical activity has also been shown to positively influence the gut microbiota, promoting a more diverse and stable microbial community. Furthermore, adequate rest and stress management are essential for microbiome health, as chronic stress can disrupt the balance of gut bacteria. By incorporating these lifestyle modifications, individuals can optimize their microbiome health and potentially reduce their risk of various diseases associated with dysbiosis.

XL. ROLE OF MICROBIOME IN NUTRACEUTICALS

Recent research has highlighted the pivotal role of the human microbiome in influencing the efficacy of nutraceuticals, a rapidly growing sector in the healthcare industry. The microbiome is a complex ecosystem of trillions of microorganisms residing in our bodies, predominantly in the gut, that play a crucial role in digestion, immune regulation, and metabolism. Studies have shown that the composition of the microbiome can significantly impact the absorption, metabolism, and bioavailability of various nutraceutical compounds. For instance, certain beneficial bacteria can enhance the conversion of dietary compounds into bioactive metabolites, amplifying the health benefits of nutraceuticals. On the other hand, an imbalanced or dysbiotic microbiome may impede the proper utilization of these compounds, leading to reduced efficacy. Understanding the intricate interplay between the microbiome and nutraceuticals is essential for optimizing their therapeutic potential and developing personalized interventions for improved health outcomes.

Microbiome-targeted Nutraceuticals

The development of microbiome-targeted nutraceuticals represents a promising avenue for personalized medicine and tailored interventions aimed at modulating the composition and function of the gut microbiota. These nutraceuticals, which encompass a variety of dietary supplements and functional foods, contain specific bioactive compounds that have been shown to selectively influence the growth and activity of beneficial bacteria in the gut. By promoting the growth of beneficial microbes while inhibiting the proliferation of harmful pathogens, microbiome-targeted nutraceuticals have the potential to restore microbial balance and improve overall health outcomes. Furthermore, the use of these targeted interventions can help mitigate dysbiosis, a condition characterized by microbial imbalance linked to various chronic diseases such as obesity, inflammatory bowel diseases, and metabolic disorders. As our understanding of the intricate interactions between the microbiome and human health continues to evolve, the development of microbiome-targeted nutraceuticals holds great promise for the future of personalized medicine.

Efficacy and Safety Considerations

Research into the human microbiome poses unique challenges when considering both efficacy and safety. The complex and dynamic nature of the microbiome requires careful consideration to ensure that interventions are not only effective but also safe for the individual. As we delve deeper into the microbiome's influence on health, it becomes increasingly apparent that a one-size-fits-all approach is not suitable. Factors such as genetics, lifestyle, environment, and diet all play a role in shaping an individual's microbiome, making personalized interventions a necessity. Moreover, interventions must be carefully monitored to prevent unintended consequences, such as the disruption of beneficial bacteria or the emergence of harmful pathogens. Thus, as we explore the potential of harnessing the microbiome for improved health outcomes, it is essential to balance efficacy with safety to ensure the best possible outcomes for individuals.

Market and Regulatory Aspects

Market and regulatory aspects play a critical role in shaping the landscape of human microbiome research. As the field continues to grow, there is increasing interest from both academia and industry, leading to a surge in investment and commercialization opportunities. This influx of funding has the potential to accelerate innovation and facilitate the translation of research findings into clinical applications. However, it also brings challenges related to intellectual property rights, data sharing, and ethical considerations. Regulatory bodies must ensure that the development of microbiome-based products adheres to rigorous standards to protect consumer safety and promote transparency. Market forces can drive the direction of research, shaping priorities and potentially influencing the dissemination of information. Therefore, a delicate balance must be struck between fostering innovation and safeguarding public trust in order to realize the full potential of the human microbiome in improving health outcomes.

XLI. MICROBIOME AND VETERINARY MEDICINE

Recent advancements in veterinary medicine have shed light on the importance of the microbiome in animal health. Just as in humans, the microbiome of animals plays a critical role in maintaining overall well-being. Studies have shown that imbalances in the microbiome can lead to various health issues in animals, ranging from digestive problems to allergic reactions. By understanding the intricate relationship between the microbiome and veterinary medicine, veterinarians can develop more effective treatment strategies for their animal patients. From probiotics to fecal transplants, the use of microbiome-based therapies in veterinary medicine is revolutionizing the way we approach animal health. By incorporating this knowledge into their practice, veterinarians can provide better care for a diverse range of species, ultimately improving the quality of life for animals under their care. As our understanding of the microbiome expands, so too will our ability to promote health and well-being in both humans and animals alike.

Applications in Animal Health

Recent advancements in microbiome research have shown promising applications in animal health. By studying the microbiomes of various animal species, scientists have gained valuable insights into how these microbial communities influence the health and well-being of their hosts. For example, research has shown that the gut microbiome plays a crucial role in the digestion and absorption of nutrients in animals, affecting their overall health and performance. Additionally, understanding the microbiome of livestock animals has led to the development of probiotics and other microbial-based therapies to improve animal health and reduce the need for antibiotics. These findings highlight the potential for microbiome-based interventions to revolutionize animal health management practices, promoting better welfare and productivity outcomes for animals across various sectors. As our understanding of the microbiome continues to expand, so do the opportunities to harness its potential for improving animal health outcomes.

Comparative Studies with Human Microbiomes

Research in the field of human microbiome has evolved significantly in recent years, shedding light on the dynamic relationship between microbial communities within our bodies and their impact on our health. Comparative studies with human microbiomes have provided valuable insights into the diversity of bacteria present in different individuals and populations, highlighting the role of genetics, diet, lifestyle, and environmental factors in shaping these microbial communities. By examining the microbiomes of various populations and comparing them across different ethnicities, geographic locations, and health statuses, researchers can identify patterns and correlations that may offer clues to understanding the complex interactions between bacteria and human health. Furthermore, comparative studies can help elucidate the mechanisms by which certain bacteria contribute to diseases or provide protective effects, paving the way for personalized medicine approaches tailored to individual microbiome compositions. Overall, comparative studies with human microbiomes hold immense potential for advancing our knowledge of microbial ecosystems within the body and their implications for health and disease.

Future Directions in Veterinary Applications

As we look towards future directions in veterinary applications, it is crucial to explore the potential benefits of incorporating microbiome research into veterinary medicine. By understanding the complex interplay between the microbiota and the host animal, veterinarians can develop novel strategies for promoting animal health and welfare. One promising avenue for future research is the development of personalized probiotics tailored to individual animals based on their unique microbiome composition. This personalized approach could revolutionize the field of veterinary medicine by providing targeted, effective treatments for a wide range of health conditions in animals. Additionally, studying the veterinary microbiome could also offer valuable insights into zoonotic diseases and antimicrobial resistance, allowing for better disease prevention and management in both animals and humans. By embracing the growing body of research on the microbiome, veterinarians can pave the way for a new era of precision medicine in animal healthcare.

XLII. MICROBIOME AND AGRICULTURAL SCIENCES

The integration of microbiome research in agricultural sciences has opened up new possibilities for enhancing crop productivity and sustainability. By exploring the complex interactions between soil microbes and plants, scientists can develop innovative strategies to improve nutrient uptake, disease resistance, and overall crop health. Understanding the role of the microbiome in agricultural ecosystems can lead to the development of biofertilizers, biopesticides, and other biological agents that can reduce the reliance on chemical inputs, promoting a more environmentally friendly approach to farming. Furthermore, harnessing the power of beneficial microbes can help mitigate the effects of climate change on agriculture, by enhancing plant resilience to extreme weather conditions. By delving deeper into the intricate relationship between microbes and plants, agricultural scientists can pave the way for a more sustainable and productive future in food production.

Impact on Soil and Plant Health

The human microbiome has a profound impact on soil and plant health through various mechanisms. Microbial populations in the gut can influence the absorption and cycling of nutrients, such as nitrogen and phosphorus, which are essential for plant growth. These microbes also play a crucial role in maintaining soil structure and fertility by aiding in the decomposition of organic matter and the formation of humus. Moreover, the bacteria present in the human body can interact with plant roots, promoting growth and increasing resistance to pathogens. This intricate relationship between the human microbiome and soil/plant health highlights the interconnectedness of all living organisms on Earth. By understanding and harnessing the power of our microbiome, we can potentially revolutionize agricultural practices and improve food security for future generations. This underscores the importance of further research in this field to maximize the benefits of this symbiotic relationship.

Applications in Sustainable Agriculture

Sustainable agriculture can greatly benefit from harnessing the potential of the human microbiome. By understanding the intricate relationships between bacteria and plants, researchers can develop innovative ways to enhance soil health, increase crop yields, and reduce the need for harmful chemical fertilizers and pesticides. For example, the application of beneficial microbes to crops can improve nutrient uptake, strengthen plant immunity, and promote overall plant growth. This not only leads to healthier crops but also contributes to environmental sustainability by minimizing the negative impacts of conventional farming practices. Furthermore, the use of microbiome-based strategies in agriculture can help reduce water usage, combat soil erosion, and protect biodiversity. By integrating knowledge of the human microbiome into sustainable agriculture practices, we can foster a more resilient and eco-friendly food production system for the future.

Future Agricultural Strategies

In light of the growing challenges posed by climate change and population growth, it is imperative to look towards future agricultural strategies that can sustainably meet the global demand for food. One promising approach is the integration of precision farming techniques, such as the use of sensor technology and data analytics, to optimize resource management and increase crop yields. By employing these cutting-edge technologies, farmers can minimize waste, reduce environmental impact, and enhance overall productivity. Additionally, investing in research and development of genetically modified crops that are resistant to pests, diseases, and adverse weather conditions could significantly improve food security worldwide. Embracing agroecological principles, such as crop rotation, intercropping, and organic farming methods, can also contribute to building resilient agricultural systems that are less reliant on synthetic inputs. By implementing a holistic approach that combines technological innovation with ecological principles, we can pave the way for a sustainable and prosperous future in agriculture.

XLIII. MICROBIOME AND FOOD INDUSTRY

The relationship between the microbiome and the food industry is a complex and multifaceted one. In recent years, there has been growing interest in understanding how the foods we consume can impact the composition and function of the microbiome in our bodies. The food industry plays a crucial role in shaping our dietary habits and thus has the potential to influence the health of our microbiome. Processed foods high in sugar, salt, and unhealthy fats have been shown to negatively impact the diversity and balance of gut bacteria, leading to a host of health issues. On the other hand, foods rich in fiber, prebiotics, and probiotics can promote the growth of beneficial bacteria and support a healthy microbiome. As such, there is a pressing need for increased collaboration between the food industry and the scientific community to develop healthier food products that can positively influence our microbiome and ultimately improve our overall health and well-being.

Influence on Food Processing

Food processing is a critical aspect of modern society, providing convenience and accessibility to a wide range of food products. The influence of the human microbiome on food processing is a topic of increasing interest and relevance. Research has shown that the microbes residing in our bodies can interact with food components during digestion, fermentation, and metabolism, impacting the overall nutritional quality of the food we consume. For example, certain gut bacteria are involved in the breakdown of complex carbohydrates, which can affect the glycemic index of foods. Additionally, the microbial composition of the gut can influence the bioavailability of nutrients, such as vitamins and minerals, from the diet. Understanding how the microbiome interacts with food processing can lead to the development of targeted dietary interventions that promote health and prevent disease. Therefore, exploring this complex relationship is crucial for advancing our knowledge of how the human microbiome influences our overall well-being.

Probiotics in Food Products

Emerging research has shown that incorporating probiotics into food products can have a positive impact on human health. Probiotics, which are live bacteria and yeasts that are beneficial for our digestive system, can help maintain a healthy balance of gut microbiota. By consuming probiotic-rich foods such as yogurt, kefir, and fermented vegetables, individuals can support their immune system, improve digestion, and even potentially reduce the risk of certain diseases. The inclusion of probiotics in food products not only provides a convenient way for individuals to boost their gut health but also opens up new avenues for innovation in the food industry. Companies are now developing a wide range of probiotic-enhanced foods, from snacks to beverages, to cater to the growing demand for functional foods that promote health and well-being. As more research is conducted in this field, the potential benefits of probiotics in food products continue to expand, offering promising prospects for improving human health in the future.

Future Trends in Food Technology

Advancements in food technology are shaping the way we consume, produce, and think about food. As we look towards the future, one trend that is gaining momentum is the use of biotechnology to enhance the nutritional value and safety of our food. Researchers are exploring the potential of genetically modified organisms (GMOs) to increase crop yield, improve resistance to pests and diseases, and boost the nutrient content of foods. With the global population projected to reach 9 billion by 2050, these innovations are crucial for ensuring food security and sustainability. Additionally, there is a growing interest in personalized nutrition, with technology enabling the customization of diets based on individual genetic makeup and microbiome composition. This targeted approach has the potential to optimize health outcomes and prevent chronic diseases. By embracing these future trends in food technology, we can revolutionize our food systems and improve the overall well-being of society.

XLIV. GLOBAL HEALTH INITIATIVES AND THE MICROBIOME

A burgeoning area of research in the realm of global health initiatives is the exploration of the human microbiome and its impact on health outcomes. The microbiome, consisting of trillions of microorganisms inhabiting the human body, has been implicated in various physiological processes, ranging from digestion to immune function. Understanding the intricate interactions between the microbiome and human health has led to innovative approaches in healthcare, such as targeted probiotics and personalized nutrition plans. By harnessing the potential of the microbiome, researchers and healthcare professionals are aiming to develop novel strategies for disease prevention and treatment on a global scale. Initiatives focused on cultivating a healthy microbiome hold promise for revolutionizing healthcare practices and promoting overall well-being across diverse populations. As the field of microbiome research continues to expand, incorporating these findings into global health initiatives is poised to have a transformative impact on public health and healthcare delivery worldwide.

International Health Programs

Recent advancements in the understanding of the human microbiome have emphasized the importance of international health programs in promoting global well-being. By considering the diverse array of microbial communities that inhabit the human body, these programs can address health disparities and diseases that transcend national borders. Utilizing a multidisciplinary approach, international health initiatives can implement strategies to improve the gut microbiome, enhance immune responses, and combat infectious diseases on a global scale. Targeted interventions, such as probiotics, personalized medicine, and public health campaigns, have the potential to reshape healthcare systems and reduce the burden of chronic illnesses worldwide. Moreover, collaborations between nations can facilitate the exchange of knowledge and resources to support research, innovation, and policy development in the field of microbiome science. Ultimately, international health programs play a vital role in promoting a holistic understanding of how bacteria in the body influence human health and pave the way for improved healthcare outcomes on a global scale.

Role of Microbiome Research in Global Health

Research on the human microbiome plays a crucial role in advancing global health initiatives by shedding light on the intricate relationship between the microorganisms within our bodies and various health outcomes. By uncovering the complexities of how these bacterial communities interact with our immune system, metabolism, and overall physiological functions, microbiome research offers valuable insights into the prevention and treatment of a myriad of health conditions. Understanding the role of the microbiome in maintaining homeostasis and protecting against pathogens has the potential to revolutionize healthcare practices worldwide. Furthermore, the integration of microbiome data into personalized medicine approaches can lead to more targeted and effective interventions tailored to individual patients, ultimately improving health outcomes on a global scale. As we continue to delve deeper into the world of the human microbiome, the implications for global health are vast and transformative.

Strategies for Global Health Improvement

The implementation of effective strategies for global health improvement is essential in addressing the complex challenges faced by populations around the world. One key strategy is investing in primary healthcare systems that focus on preventive measures and early intervention, rather than just disease treatment. This approach can help reduce the burden on healthcare systems and improve overall health outcomes. Additionally, promoting education and awareness about proper hygiene practices, nutrition, and vaccination programs can have a significant impact on preventing the spread of infectious diseases and improving overall health. Furthermore, collaborating with international organizations, governments, and local communities to develop sustainable health solutions tailored to specific regions can ensure long-term success in promoting global health equity. By integrating these strategies, we can work towards a healthier world for all individuals, regardless of geographical location or socioeconomic status.

XLV. MICROBIOME AND BIOTECHNOLOGY

Biotechnology has opened up exciting possibilities for utilizing the human microbiome in various applications. By harnessing the power of these microorganisms, researchers can develop innovative solutions for healthcare, agriculture, and environmental remediation. One promising avenue is the development of probiotics, live bacteria that can confer health benefits when consumed in adequate amounts. These probiotics can be tailored to target specific conditions, such as inflammatory bowel disease or metabolic disorders. Additionally, the microbiome can be leveraged in the production of biofuels and bioplastics, offering sustainable alternatives to traditional petrochemical-based products. With a better understanding of the intricate interactions between the microbiome and biotechnology, the potential for advancements in medicine, agriculture, and environmental science is vast. As we continue to explore this burgeoning field, the integration of microbiome-based solutions into various industries holds promise for improving human health and sustainability.

Biotechnological Applications

The applications of biotechnology in understanding the human microbiome have opened up a new frontier in health research. By leveraging biotechnological tools such as next-generation sequencing, scientists can now delve deep into the complex microbial communities that inhabit our bodies. This has allowed for the identification of specific bacterial species that play key roles in processes like digestion, immune response, and even mental health. Furthermore, biotechnological advancements have paved the way for the development of personalized medicine based on an individual's unique microbiome profile. This targeted approach holds great promise for treating conditions ranging from gastrointestinal disorders to autoimmune diseases. As our understanding of the human microbiome continues to grow, so too does the potential for innovative biotechnological interventions that could revolutionize healthcare practices and improve overall well-being.

Innovations in Microbiome Engineering
Recent advancements in microbiome engineering have opened up new possibilities for understanding and manipulating the complex microbial communities that inhabit our bodies. By utilizing cutting-edge technologies such as CRISPR-Cas9 and synthetic biology, researchers can target specific bacteria within the microbiome to modify their genetic makeup, potentially leading to novel treatments for a variety of health conditions. For instance, engineered probiotics could be designed to produce therapeutic molecules or to selectively target pathogenic microbes, offering a more precise and personalized approach to combating infections or diseases. Additionally, bioengineered microbial consortia may be able to restore microbial balance in disrupted microbiomes, promoting overall health and well-being. These innovative strategies in microbiome engineering hold great promise for revolutionizing the field of medicine and improving human health in ways previously thought impossible. By harnessing the power of microbial manipulation, we may unlock a new era of personalized medicine tailored to the individual's unique microbiome profile.

Ethical and Safety Considerations

Furthermore, ethical and safety considerations must be carefully examined when conducting research on the human microbiome. As we delve deeper into understanding the intricate relationship between the bacteria in our bodies and our health, it is essential to prioritize the well-being and rights of the individuals involved. Researchers must ensure that informed consent is obtained from participants, and that their autonomy and privacy are respected throughout the study. Additionally, measures should be taken to minimize any potential risks or harm that may arise from microbiome-related interventions or treatments. This involves thorough monitoring of potential side effects and adverse reactions, as well as adherence to strict ethical guidelines to ensure the safety of participants. By upholding these ethical principles and prioritizing safety considerations, researchers can conduct meaningful and responsible investigations into the human microbiome while upholding the highest standards of integrity and respect for human subjects.

XLVI. CHALLENGES IN TRANSLATING MICROBIOME RESEARCH

Research on the human microbiome has revealed a plethora of challenges when it comes to translating findings into practical applications. One major hurdle is the sheer complexity and diversity of the microbiome itself. With trillions of microorganisms residing in the human body, identifying and understanding their individual roles and interactions can be a monumental task. Moreover, the microbiome is highly dynamic, influenced by various factors such as diet, environment, and genetics. These complexities make it difficult to draw definitive conclusions from research findings and apply them in a clinical setting. Additionally, there is a lack of standardized methodologies and tools for studying the microbiome, leading to inconsistencies in research outcomes. As we strive to bridge the gap between microbiome research and real-world applications, addressing these challenges will be essential to harnessing the full potential of this burgeoning field for improving human health.

From Laboratory to Clinic

The transition from laboratory to clinic in the study of the human microbiome marks a critical junction where theoretical knowledge is translated into practical applications for the benefit of human health. As researchers uncover the intricate interactions between the microbiome and various disease states, the potential for targeted interventions and personalized treatments becomes increasingly feasible. By identifying specific bacterial populations associated with certain conditions, such as inflammatory bowel disease or obesity, clinicians can develop tailored therapies that aim to restore microbial balance and promote wellness. Furthermore, advancements in technology, such as metagenomics and bioinformatics, have enabled a more comprehensive understanding of the microbiome's role in human physiology. As we bridge the gap between research findings and clinical practice, the integration of microbiome-based approaches into mainstream healthcare holds promise for revolutionizing how we diagnose, treat, and prevent disease in the future.

Barriers in Clinical Application
The integration of research findings from the human microbiome into clinical practice encounters various barriers that hinder its effective application. One significant challenge is the complexity and diversity of the microbiota, making it difficult to identify consistent patterns or establish clear cause-and-effect relationships between microbial compositions and health outcomes. Additionally, the lack of standardized methodologies for sampling, analysis, and interpretation of microbiome data poses a substantial obstacle in translating research findings into actionable clinical interventions. Moreover, the absence of comprehensive guidelines or protocols for incorporating microbiome-based diagnostics and therapeutics into existing healthcare systems further impedes the integration of this emerging field into routine clinical practice. To overcome these barriers, efforts should be directed towards developing standardized protocols, fostering interdisciplinary collaborations, and promoting evidence-based guidelines to support the effective application of microbiome research in clinical settings.

Strategies for Overcoming Challenges
In order to address the myriad challenges associated with understanding and manipulating the human microbiome, various strategies can be employed. One key approach is to leverage advanced sequencing technologies to unravel the complex interactions between microbial communities and host cells. This can provide valuable insights into the composition and functions of the microbiome, shedding light on how it contributes to health or disease. Additionally, targeted interventions such as probiotics, prebiotics, and fecal microbiota transplantation offer promising avenues for modulating the microbiome and restoring microbial balance in the body. Moreover, fostering interdisciplinary collaborations between researchers from different fields – such as microbiology, immunology, and bioinformatics – can lead to innovative solutions for deciphering the complexities of the microbiome. By combining these strategies, researchers can overcome the challenges posed by the human microbiome and pave the way for novel therapeutic interventions that harness the power of beneficial bacteria to promote health and prevent disease.

XLVII. MICROBIOME AND PUBLIC SAFETY

The intricate relationship between the human microbiome and public safety is an emerging area of research that carries significant implications for society. As we delve deeper into the microbial communities residing within our bodies, we unveil a complex network of interactions that extend beyond individual health outcomes. The microbiome has the potential to shape public safety through its involvement in various infectious diseases, antibiotic resistance, and even mental health disorders. Understanding how the microbiome influences these factors can lead to more effective strategies for controlling disease outbreaks, combating drug-resistant pathogens, and promoting overall well-being on a larger scale. By harnessing the power of microbiome research, we can unlock new avenues for safeguarding public health and enhancing safety measures across diverse populations. This intersection between the microbiome and public safety represents a promising frontier in healthcare and public policy, highlighting the vital role of microbial communities in shaping the welfare of society as a whole.

Biosecurity Concerns
The human microbiome poses significant biosecurity concerns that must be carefully considered in the realm of public health and medical practice. With the rise of antibiotic-resistant bacteria and the potential for pathogens to spread rapidly within populations, understanding and managing the microbiome is essential. Research has shown that disruptions in the normal composition of the microbiome can lead to a host of health issues, from inflammatory disorders to metabolic imbalances.

Therefore, it is crucial to develop strategies for maintaining a healthy microbiome and mitigating the risks associated with harmful microbial communities. By promoting a balanced ecosystem of beneficial bacteria, we can potentially reduce the incidence of infections and improve overall well-being. Harnessing the power of the microbiome while being mindful of biosecurity risks can pave the way for innovative approaches to healthcare that prioritize prevention and personalized treatments.

Microbiome in Disease Surveillance

Emerging evidence suggests a significant role for the human microbiome in disease surveillance. By analyzing the composition and dynamics of the microbiota, researchers can potentially identify early warning signs of various diseases, enabling preemptive interventions before clinical symptoms manifest. This approach holds immense promise in preventing the onset of conditions such as inflammatory bowel disease, diabetes, and even certain types of cancer. Furthermore, the microbiome's ability to influence the immune system and metabolic processes underscores its importance as a key player in disease development and progression. Leveraging this knowledge for personalized medicine could revolutionize healthcare by enabling targeted interventions that address the root cause of illnesses rather than merely treating symptoms. As research in this field continues to advance, integrating microbiome analysis into disease surveillance protocols has the potential to enhance early detection, improve treatment outcomes, and ultimately redefine the paradigm of healthcare delivery.

Public Safety Strategies

Moreover, in order to maintain public safety, it is imperative to implement a range of strategies that address the complex nature of modern threats. One key approach is the use of technology in enhancing surveillance and monitoring systems, allowing for early detection of potential risks and rapid response to emergencies. Additionally, fostering community partnerships and engaging with local stakeholders can help build trust, promote information sharing, and increase overall resilience. Training programs for law enforcement personnel and first responders are essential in ensuring a proactive and effective response to crises. Furthermore, investing in research and development to stay ahead of emerging threats and continuously reassessing and updating public safety policies are crucial components of a comprehensive strategy. By embracing a multidimensional approach that combines technology, community engagement, training, and innovation, we can work towards a safer and more secure future for all.

XLVIII. FUTURE DIRECTIONS IN MICROBIOME RESEARCH

As we look to the future of microbiome research, it is essential to consider the potential applications of our growing knowledge in this field. One promising avenue for exploration is the development of personalized microbiome-based treatments tailored to individual patients. By understanding how specific bacteria influence health outcomes in different individuals, we can work towards more targeted interventions that take into account the unique microbiome composition of each person. Additionally, further research is needed to elucidate the mechanisms through which the microbiome influences various diseases, paving the way for novel therapeutic approaches. Integrating advances in microbiome research with other fields such as genetics and immunology will also be crucial in unraveling the complex interactions within the human body. Moving forward, collaboration between researchers from diverse disciplines will be key to unlocking the full potential of microbiome-based interventions for enhancing human health and well-being.

Emerging Research Areas

As research in the field of human microbiome continues to expand, several emerging areas are gaining traction in the scientific community. One such area of interest is the role of the gut-brain axis in influencing mental health and cognitive function. By studying the intricate relationship between the gut microbiota and the central nervous system, researchers hope to uncover new insights into the mechanisms by which our gut bacteria can impact our mood, behavior, and even neurological diseases. Additionally, advancements in metagenomics and computational biology have enabled scientists to delve deeper into the complex microbial communities residing in various body sites, leading to a better understanding of the overall impact of the microbiome on human health. Through these emerging research areas, we are beginning to unravel the profound influence that bacteria in our bodies have on our well-being, paving the way for innovative therapeutic interventions and personalized medicine approaches in the future.

Potential Breakthroughs

Thus far, research into the human microbiome has revealed countless potential breakthroughs that could revolutionize healthcare in the future. One exciting area of discovery lies in the relationship between gut bacteria and mental health. Studies have shown that the microbiota in our intestines may influence our mood, behavior, and even cognitive function through the gut-brain axis. This opens up new possibilities for treating mental health disorders such as depression and anxiety by targeting the gut microbiome. Another promising avenue of research is the potential for personalized microbiome-based therapies. By understanding each individual's unique microbiota composition, scientists may be able to develop personalized treatments that could target specific health issues more effectively. These breakthroughs have the potential to vastly improve the quality of healthcare and revolutionize our understanding of the intricate connection between our bodies and the trillions of bacteria that reside within us.

Long-term Research Goals

As we look towards the future of human microbiome research, it is essential to establish long-term goals that will guide our efforts in understanding the intricate relationship between the bacteria within our bodies and our health. One of the primary objectives should be to explore the potential therapeutic applications of manipulating the microbiome to treat a variety of diseases and conditions. This includes developing targeted interventions that can restore microbial balance in individuals suffering from dysbiosis or other microbiome-related disorders. Additionally, long-term research goals should focus on elucidating the mechanisms by which the microbiome influences immune function, metabolism, and neurological processes. By unraveling these intricate pathways, we can uncover new opportunities for interventions that harness the power of the microbiome to promote health and prevent disease. Ultimately, establishing clear long-term research goals will ensure that we continue to make significant strides in unlocking the full potential of the human microbiome.

XLIX. SUMMARY OF KEY FINDINGS

In examining the key findings of the human microbiome, it becomes evident that the intricate relationship between the bacteria residing within our bodies and our overall health is a multifaceted and dynamic one. Through extensive research and analysis, it has been established that the composition of the microbiome can have profound implications for various aspects of human health, ranging from metabolism and immune function to mental well-being and disease susceptibility. The diversity and stability of the microbiome have emerged as crucial factors in maintaining a balanced and harmonious relationship between the microorganisms and the host. Moreover, the influence of external factors such as diet, lifestyle, and environmental exposures on the composition of the microbiome further highlights the intricate interplay between the microbial communities and human physiology. These key findings underscore the importance of understanding and nurturing the human microbiome to support optimal health and well-being.

Major Insights from the Essay

The essay on the human microbiome offers several major insights into the intricate relationship between our bodies and the bacteria that inhabit them. One key revelation is the significant impact these microorganisms have on our health, influencing everything from our immune system to our metabolism. By understanding the delicate balance of the microbiome, researchers can develop targeted therapies to treat a myriad of diseases and conditions. Additionally, the essay sheds light on the importance of maintaining a diverse and healthy microbiome

through factors such as diet, lifestyle choices, and medical interventions. This emphasizes the need for personalized approaches to healthcare that take into account an individual's unique microbiome profile. Overall, the essay underscores
the vital role that bacteria play in shaping our health and highlights the potential for harnessing this knowledge to improve outcomes in various health conditions.

Implications for Future Research

Moving forward, the implications for future research in understanding the human microbiome are vast and promising. Research has already shown the intricate relationship between the microbiome and various health conditions, opening up avenues for targeted therapies and interventions. Future studies could delve deeper into the mechanisms by which specific bacterial strains influence metabolic processes, immune responses, and overall health outcomes. Additionally, exploring the role of the microbiome in different populations and how it may impact personalized medicine could lead to more tailored interventions for individuals based on their unique microbial composition. Further research into the dynamic nature of the microbiome and how it responds to lifestyle changes, diet, and external factors could provide insights into preventative strategies and treatment options. As we continue to unravel the complexities of the human microbiome, the possibilities for improving health outcomes are immense.

Relevance to Health and Disease

The relevance of the human microbiome to health and disease cannot be overstated. Research has shown that the trillions of bacteria living in and on our bodies have a direct impact on our immune system, metabolism, and even our mental health. Dysbiosis, or an imbalance in the microbiome, has been linked to a range of health issues, including obesity, diabetes, inflammatory bowel disease, and even certain types of cancer. Understanding the interplay between the microbiome and disease is essential for developing personalized treatments and interventions that target the root cause of health issues rather than just treating symptoms. By focusing on restoring balance to the microbiome through probiotics, prebiotics, or fecal transplants, researchers and medical professionals can potentially revolutionize the way we approach and treat a wide range of health conditions. The implications of microbiome research on human health are profound, offering new avenues for therapeutic interventions and preventive strategies that could improve the lives of millions worldwide.

L. IMPLICATIONS FOR POLICY AND PRACTICE

The understanding of the human microbiome has profound implications for policy and practice in the realms of healthcare, nutrition, and public health. By recognizing the intricate relationship between the microbes residing in our bodies and our overall well-being, policymakers can formulate targeted interventions to prevent and treat a multitude of health conditions. For instance, initiatives promoting a diverse and balanced microbiome through diet, probiotics, or fecal transplants could revolutionize the way we approach chronic diseases like obesity, autoimmune disorders, and mental health issues. Furthermore, integrating microbiome analysis into routine medical practice could enable personalized treatment plans that harness the power of beneficial bacteria to optimize individual health outcomes. By bridging the gap between scientific research and practical applications, advancements in microbiome science have the potential to transform healthcare delivery and shape public health policies for years to come.

Recommendations for Health Practitioners
In light of the significant impact that the human microbiome has on our health, it is imperative that health practitioners are well-informed about this intricate system within the body. One key recommendation for health practitioners is to stay updated on the latest research and findings in microbiome science. Understanding the role of different bacterial species and their interactions can lead to more personalized and effective treatment

plans for patients. Additionally, health practitioners should consider the use of probiotics and prebiotics in promoting a healthy microbiome, as these supplements can help restore balance in the gut microbiota. Furthermore, educating patients about the importance of maintaining a diverse and resilient microbiome through a balanced diet rich in fiber and fermented foods is essential. By incorporating these recommendations into their practice, health practitioners can enhance patient outcomes and improve overall health and well-being.

Policy Implications

The implications of understanding the human microbiome extend far beyond the realms of healthcare and medicine. Policymakers must consider the broader societal impact of microbiome research, particularly in areas such as public health, food production, and environmental sustainability. One critical policy implication is the need for increased funding and support for microbiome research to unlock its full potential in revolutionizing personalized medicine and preventive health strategies. Additionally, policymakers must address issues related to public awareness and education about the importance of a healthy microbiome, as well as regulations on the use of microbiome-based therapies and products. Furthermore, policies should be put in place to promote the development of microbiome-friendly environments, both in healthcare settings and in everyday life. By incorporating microbiome research into policy decisions, we can pave the way for a healthier future for all individuals.

Practical Applications of Research Findings

The practical applications of research findings related to the human microbiome are vast and offer great potential for improving health outcomes. One area where these findings can be applied is in personalized medicine, where the unique composition of an individual's microbiome can be used to tailor treatment plans for various health conditions. By understanding how specific bacteria interact with the body, researchers can develop targeted therapies that work in harmony with the microbiome rather than against it. Furthermore, research on the human microbiome has the potential to revolutionize the fields of nutrition and probiotics, as scientists uncover the role that gut bacteria play in digestion, metabolism, and overall health. Ultimately, the practical applications of research on the human microbiome have the potential to transform healthcare practices and improve outcomes for a wide range of individuals.

LI. CONCLUSION

In conclusion, the human microbiome is a complex and dynamic ecosystem that plays a significant role in our health and well-being. Through interactions with our immune system, metabolism, and brain, the trillions of bacteria residing in our bodies exert profound effects on various aspects of our physiology. Understanding the intricacies of the human microbiome opens up new avenues for therapeutic interventions and personalized medicine. By modulating the composition of our microbial communities through diet, probiotics, or fecal transplants, we may be able to improve various health conditions ranging from gastrointestinal disorders to mental health issues. However, further research is needed to fully elucidate the mechanisms underlying the microbiome's influence on human health and to develop targeted interventions that harness the potential of these microscopic allies. As we continue to unravel the mysteries of the human microbiome, we are entering an exciting era of medical science where precision healthcare tailored to our unique microbial profiles may become a reality.

Recapitulation of Thesis and Main Points
In conclusion, the analysis of the human microbiome has revealed the intricate relationship between the bacteria residing in our bodies and their impact on our health. Through the exploration of various research studies and findings, it has become evident that the microbiome plays a significant role in regulating immune responses, metabolism, and even mental health. By understanding the delicate balance within our microbial commu-

nities, we can potentially harness this knowledge to develop targeted interventions for a range of health conditions. Furthermore, the diversity and composition of the microbiome have been shown to vary significantly among individuals, highlighting the need for personalized approaches in healthcare. As we continue to unravel the complexities of the human microbiome, it is imperative that we recognize its potential as a key player in maintaining optimal health and well-being. The implications of this research are vast and have the potential to revolutionize the way we approach healthcare in the future.

Future Outlook in Microbiome Research

As we look towards the future of microbiome research, significant advancements and discoveries are on the horizon. With emerging technologies such as metagenomics, metabolomics, and single-cell sequencing, researchers are gaining unprecedented insights into the intricate relationship between the microbiome and human health. These tools allow for a deeper understanding of the complex interactions within the microbiome ecosystem and its impact on various disease states. Furthermore, the integration of artificial intelligence and machine learning algorithms in analyzing vast amounts of microbiome data holds the potential to uncover novel therapeutic interventions and personalized treatments. As we continue to unravel the complexities of the human microbiome, the future is promising for the development of targeted therapies that harness the power of our microbial inhabitants to improve health outcomes and reshape the landscape of medicine. The potential for microbiome research to revolutionize healthcare is vast, offering a new frontier of possibilities in personalized medicine and disease prevention.

Closing Remarks

In conclusion, the human microbiome is a complex ecosystem of bacteria that play a vital role in maintaining our health and influencing our susceptibility to disease. The interplay between the microbes that inhabit our bodies and our overall well-being is a dynamic field of study that continues to reveal new insights into the intricacies of human biology. By understanding the delicate balance of the microbiome, we can potentially unlock new avenues for therapeutic interventions and personalized medicine. As research in this field progresses, it is becoming increasingly clear that our microbial inhabitants are not mere bystanders but active participants in shaping our health outcomes. Moving forward, it will be crucial to further explore the interactions between our microbiome and various disease states to develop targeted interventions that harness the power of these microscopic allies for the betterment of human health.

BIBLIOGRAPHY

Barry White. 'Mapping Your Thesis.' ACER Press, 6/1/2011

Committee on Assessing Rehabilitation Science and Engineering. 'Enabling America.' Assessing the Role of Rehabilitation Science and Engineering, Institute of Medicine, National Academies Press, 11/24/1997

National Academy of Sciences. 'Policy Implications of Greenhouse Warming.' Mitigation, Adaptation, and the Science Base, National Academy of Engineering, National Academies Press, 2/1/1992

World Health Organization. 'Improving Healthcare Quality in Europe Characteristics, Effectiveness and Implementation of Different Strategies.' Characteristics, Effectiveness and Implementation of Different Strategies, OECD, OECD Publishing, 10/17/2019

Jeffrey J. Shook. 'Childhood, Youth, and Social Work in Transformation.' Implications for Policy and Practice, Lynn M. Nybell, Columbia University Press, 1/30/2009

Division of Behavioral and Social Sciences and Education. 'Knowing What Students Know.' The Science and Design of Educational Assessment, National Research Council, National Academies Press, 10/27/2001

Santosh Kumar. 'Component-Based Software Engineering.' Methods and Metrics, Umesh Kumar Tiwari, CRC Press, 11/18/2020

United States. Congress. House. Committee on Science and Technology. Subcommittee on Natural Resources, Agriculture Research, and Environment. 'The Nation's Long-term Agriculture Research Needs.' Hearings Before the Subcommittee on Natural Resources, Agriculture Research, and Environment of the Committee on Science and Technology, U.S. House of Representatives, Ninety-seventh Congress, Second Session, July 27, 29, 1982, U.S. Government Printing Office, 1/1/1983

Andrew Hargadon. 'How Breakthroughs Happen.' The Surprising Truth about how Companies Innovate, Harvard Business Press, 1/1/2003

Division on Earth and Life Studies. 'New Research Directions for the National Geospatial-Intelligence Agency.' Workshop Report, National Research Council, National Academies Press, 8/18/2010

P.J. Ortmeier. 'Public Safety and Security Administration.' Gulf Professional Publishing, 9/10/1998

Board on Life Sciences. 'Biosecurity Challenges of the Global Expansion of High-Containment Biological Laboratories.' Committee on Anticipating Biosecurity Challenges of the Global Expansion of High-Containment Biological Laboratories, National Academies Press, 3/16/2012

Adriano Fabris. 'Trust.' A Philosophical Approach, Springer Nature, 4/6/2020

Committee on the Robert Wood Johnson Foundation Initiative on the Future of Nursing, at the Institute of Medicine. 'The Future of Nursing.' Leading Change, Advancing Health, Institute of Medicine, National Academies Press, 2/8/2011

Jorge L. Sepulveda. 'Accurate Results in the Clinical Laboratory.' A Guide to Error Detection and Correction, Amitava Dasgupta, Newnes, 1/22/2013

American Nurses Association. 'Code of Ethics for Nurses with Interpretive Statements.' Nursesbooks.org, 1/1/2001

Matthew W. Chang. 'Principles in Microbiome Engineering.' John Wiley & Sons, 5/3/2022

James C. Ogbonna. 'Microbiomes and Emerging Applications.' Nwadiuto (Diuto) Esiobu, CRC Press, 5/10/2022

Robert E. Black. 'Global Health.' Diseases, Programs, Systems, and Policies, Michael H. Merson, Jones & Bartlett Publishers, 8/19/2011

Ram B. Singh. 'The Role of Functional Food Security in Global Health.' Ronald Ross Watson, Elsevier Science, 11/13/2018

Craig Leadley. 'Innovation and Future Trends in Food Manufacturing and Supply Chain Technologies.' Woodhead Publishing, 11/18/2015

Sarah Wernick. 'The Probiotics Revolution.' The Definitive Guide to Safe, Natural Health Solutions Using Probiotic and Prebiotic Foods and Supplements, Gary B. Huffnagle, Random House Publishing Group, 6/24/2008

Marion Nestle. 'Food Politics.' How the Food Industry Influences Nutrition and Health, University of California Press, 5/14/2013

Roland R. Arnold. 'How Fermented Foods Feed a Healthy Gut Microbiota.' A Nutrition Continuum, M. Andrea Azcarate-Peril, Springer Nature, 11/28/2019

Thomas A. Lyson. 'Remaking the North American Food System.' Strategies for Sustainability, C. Clare Hinrichs, U of Nebraska Press, 1/1/2007

Alexander Wezel. 'Agroecological Practices For Sustainable Agriculture: Principles, Applications, And Making The Transition.' World Scientific, 6/19/2017

Helen Hayden. 'Soil Health, Soil Biology, Soilborne Diseases and Sustainable Agriculture.' A Guide, Graham Stirling, Csiro Publishing, 3/1/2016

Saima Hamid. 'Microbiomics and Sustainable Crop Production.' Mohammad Yaseen Mir, John Wiley & Sons, 3/27/2023

Division on Earth and Life Studies. 'Environmental Chemicals, the Human Microbiome, and Health Risk.' A Research Strategy, National Academies of Sciences, Engineering, and Medicine, National Academies Press, 3/1/2018

Arlene McDowell. 'Long Acting Animal Health Drug Products.' Fundamentals and Applications, Michael J. Rathbone, Springer Science & Business Media, 10/12/2012

Glenn Zhang. 'Gut Microbiota, Immunity, and Health in Production Animals.' Michael H. Kogut, Springer Nature, 1/19/2022

Ezekiel J. Emanuel. 'Ethical and Regulatory Aspects of Clinical Research.' Readings and Commentary, Johns Hopkins University Press, 1/1/2003

Sarvadaman Pathak. 'Anxiety, Gut Microbiome, and Nutraceuticals.' Recent Trends and Clinical Evidence, Yashwant V. Pathak, CRC Press, 9/26/2023

Alan C. Logan. 'The Secret Life of Your Microbiome.' Why Nature and Biodiversity are Essential to Health and Happiness, Susan L. Prescott, New Society Publishers, 9/1/2017

John C. Coffee. 'Entrepreneurial Litigation.' Its Rise, Fall, and Future, Harvard University Press, 6/8/2015

Roger Fisher. 'Improving Compliance with International Law.' University Press of Virginia, 1/1/1981

Robert P. Benko. 'Protecting Intellectual Property Rights.' Issues and Controversies, American Enterprise Institute for Public Policy Research, 1/1/1987

Laura Bowater. 'The Microbes Fight Back.' Antibiotic Resistance, Royal Society of Chemistry, 10/25/2017

Division of Health Care Services. 'The Future of Public Health.' Committee for the Study of the Future of Public Health, National Academies Press, 1/15/1988

Ralf Junker. 'Point-of-care testing.' Principles and Clinical Applications, Peter Luppa, Springer, 7/18/2018

Division of Behavioral and Social Sciences and Education. 'Fostering Healthy Mental, Emotional, and Behavioral Development in Children and Youth.' A National Agenda, National Academies of Sciences, Engineering, and Medicine, National Academies Press, 1/18/2020

B Reusens. 'Functional Foods, Ageing and Degenerative Disease.' C Remacle, Elsevier, 6/9/2004

Antara Banerjee. 'Gut Microbiome and Brain Ageing.' Brain Aging, Surajit Pathak, Springer Nature Singapore, 4/19/2024

Division on Earth and Life Studies. 'The Chemistry of Microbiomes.' Proceedings of a Seminar Series, National Academies of Sciences, Engineering, and Medicine, National Academies Press, 7/19/2017

Catherine Stanton. 'The Gut-Brain Axis.' Dietary, Probiotic, and Prebiotic Interventions on the Microbiota, Niall Hyland, Elsevier, 12/7/2023

B.E. Leonard. 'Microbes and the Mind.' The Impact of the Microbiome on Mental Health, C.S.M. Cowan, Karger Medical and Scientific Publishers, 5/6/2021

Heinz Rupp. 'Pathophysiology of Cardiovascular Disease.' Naranjan S. Dhalla, Springer Science & Business Media, 12/6/2012

Amedeo Amedei. 'Gut Microbiota and Inflammation: Relevance in Cancer and Cardiovascular Disease.' Cinzia Parolini, Frontiers Media SA, 2/9/2021

Antonio Salgado-Somoza. 'Emerging Roles of the Gut Microbiota in the Pathogenesis of Metabolic Disorders.' Isabel Moreno-Indias, Frontiers Media SA, 10/1/2021

Ramanan Laxminarayan. 'Disease Control Priorities, Third Edition (Volume 2).' Reproductive, Maternal, Newborn, and Child Health, Robert Black, World Bank Publications, 4/11/2016

Fereidoon Shahidi. 'Tree Nuts.' Composition, Phytochemicals, and Health Effects, Cesarettin Alasalvar, CRC Press, 12/17/2008

A. Lenore Ackerman. 'The Urogenital Microbiota in Urinary Tract Diseases.' Frontiers Media SA, 1/10/2023

Abigail Lois Coughtrie. 'Epidemiology and Ecology of Microbial Communities of the Upper Respiratory Tract.' Original typescript, 1/1/2015

Stavros Garantziotis. 'The Microbiome in Respiratory Disease.' Principles, Tools and Applications, Yvonne J. Huang, Springer Nature, 1/1/2022

Connie Chenevert Mobley. 'Prevention in Clinical Oral Health Care.' David P. Cappelli, Elsevier Health Sciences, 10/26/2007

Nezar Al-Hebshi. 'The Human Microbiome and Cancer.' Gary Moran, Frontiers Media SA, 9/3/2020

Roopal V. Kundu. 'Clinical Cases in Skin of Color.' Adnexal, Inflammation, Infections, and Pigmentary Disorders, Porcia B. Love, Springer, 11/6/2015

Theodore A. Sundstrom. 'Mathematical Reasoning.' Writing and Proof, Pearson Prentice Hall, 1/1/2007

Nava Dayan. 'Skin Microbiome Handbook.' From Basic Research to Product Development, John Wiley & Sons, 9/1/2020

Yehuda Shoenfeld. 'Mosaic of Autoimmunity.' The Novel Factors of Autoimmune Diseases, Carlo Perricone, Elsevier, 2/15/2019

W.A. Walker. 'Milk, Mucosal Immunity and the Microbiome: Impact on the Neonate.' P.L. Ogra, Karger Medical and Scientific Publishers, 4/24/2020

Paul Travers. 'Janeway's Immunobiology.' Kenneth Murphy, Taylor & Francis Group, 6/22/2010

Amedeo Amedei. 'The Interplay of Microbiome and Immune Response in Health and Diseases.' Gwendolyn Barcel´o-Coblijn, MDPI, 11/6/2019

Stanley B. Benjamin. 'Gastrointestinal Disease.' An Endoscopic Approach, Anthony J. DiMarino, SLACK Incorporated, 1/1/2002

Morris Green. 'The Role of the Gastrointestinal Tract in Nutrient Delivery.' The Role of the Gastrointestinal Tract in Nutrient Delivery, Academic Press, 12/2/2012

Bernard William Downs. 'Microbiome, Immunity, Digestive Health and Nutrition.' Epidemiology, Pathophysiology, Prevention and Treatment, Debasis Bagchi, Academic Press, 7/21/2022

Shailza Singh. 'Metagenomic Systems Biology.' Integrative Analysis of the Microbiome, Springer Nature, 12/7/2020

Sun Kim. 'Genome Sequencing Technology and Algorithms.' Artech House, 1/1/2008

Marc Paye. 'Handbook of Cosmetic Science and Technology.' André O. Barel, CRC Press, 4/9/2014

Antonio Mendex-Vilas. 'Microbes in Applied Research.' Current Advances and Challenges, World Scientific, 1/1/2012

Bruce Stutz. 'Theories for Everything.' An Illustrated History of Science from the Invention of Numbers to String Theory, John Langone, National Geographic Books, 1/1/2006

Food and Nutrition Board. 'The Human Microbiome, Diet, and Health.' Workshop Summary, Food Forum, National Academies Press, 2/27/2013

Alistair McCleery. 'An Introduction to Book History.' David Finkelstein, Routledge, 3/13/2006

www.ingramcontent.com/pod-product-compliance
Lightning Source LLC
Chambersburg PA
CBHW050216230526
45470CB00001B/410